INTRODUCTION TO HOLOMORPHY

NORTH-HOLLAND MATHEMATICS STUDIES 106
Notas de Matemática (98)

NORTH-HOLLAND – AMSTERDAM • NEW YORK • OXFORD

INTRODUCTION TO HOLOMORPHY

Jorge Alberto BARROSO
Universidade Federal do Rio de Janeiro
Rio de Janeiro
Brasil

1985

NORTH-HOLLAND – AMSTERDAM • NEW YORK • OXFORD

ISBN: 0444 87666 9

Publishers:
ELSEVIER SCIENCE PUBLISHERS B.V.
P.O. Box 1991
1000 BZ Amsterdam
The Netherlands

Sole distributors for the U.S.A. and Canada:
ELESEVIER SCIENCE PUBLISHING COMPANY, INC.
52 Vanderbilt Avenue
New York, N.Y. 10017
U.S.A.

Library of Congress Cataloging in Publication Data

Barroso, Jorge Alberto.
Introduction to holomorphy.

(North-Holland mathematic studies ; 106) (Notas de matematica ; 98)
Bibliography: p.
Includes index.
1. Normed linear spaces. 2. Domains of holomorphy.
I. Title. II. Series. III. Series: Notas de matemática (Amsterdam, Netherlands) ; 98.
QA1.N86 no.98 [QA322.2] 510 s [515.7'3] 84-22283
ISBN 0-444-87666-9

PRINTED IN THE NETHERLANDS

To Anna Amalia

with love.

FOREWORD

This book presents, on the one hand, a set of basic properties of holomorphic mappings between complex normed spaces and between complex locally convex spaces. These properties have already achieved an almost definitive form and should be known to all those interested in the study of infinite dimensional Holomorphy and its applications. On the other hand, for reasons of personal taste but also (and especially) because of the importance of the matter, some incursions have been made into the study of the topological properties of the spaces of holomorphic mappings between spaces of infinite dimension. An attempt is then made to show some of the several topologies that can naturally be considered in these spaces.

There has been no concern to establish priorities and relatively few authors are quoted in the text. Some historical facts should be pointed out here. The study of differential mapping and holomorphic mapping between spaces of infinite dimension apparently begins with V. Volterra [142], [143], [144], [145], [146] around 1887. Then D. Hilbert, in his work [56], outlines a theory of holomorphic mappings in an infinity of variables, in which the concept of polynomial in such a context already clearly appears. At the same time (1909), M. Fréchet publishes his first work [40] on the abstract theory of polynomials in an infinity of variables. Later on, the development of the theory of normed spaces led Fréchet to define([41],[42]) real polynomial in a more general situation. Mention must be made of R. Gâteaux's works [43], [44], in which he proposes a definition for complex polynomials. In the period until the mid-60s, several other names are worthy of note. A historical vision of the development of the notions of polynomial and holomorphic mapping in this period can be obtained through the works of A.E. Taylor, [135], [136]. The mid-60s witnessed a rekindling of interest and a quickening of the development of the study of questions that originate in the notion of holomorphic mapping between complex normed spaces and between complex locally convex spaces. Thus, infinite dimensional Holomorphy appears as a theory rich in fascinating problems and rich in applications to other branches of Mathematics and Mathematical Physics. Once again without any desire to establish priorities, we would like to quote the names of H. Cartan and P. Lelong in France, for their influence and work, as well as the French team composed of G. Coeuré, J.-F. Colombeau, A. Douady, M. Hervé, A. Hirschowitz, P. Krée, P. Mazet, P. Noverraz, P. Raboin, J.P. Ramis and J.P. Vigué. Still, we should especially like to stress the important role played in the development of this theory by Leopoldo Nachbin and his doctoral students in Brazil and the United States: A.J. Aragona, R.M. Aron, R.R. Baldino, J.A. Barrroso, P.D. Berner, P.J. Boland, S.B. Chae, S. Dineen, C.P. Gupta, G.I. Katz, M.C. Matos, J. Mujica, D.P. Pombo, R.L. Soraggi, J.O. Stevenson, A.J.M. Wanderley as well as T. Abuabara, T.A.W. Dwyer, J.M. Isidro, L.A. Moraes and D. Pisanelli, all of whom were directly influenced by him.

Mention should also be made of the German school, as represented by K.-D. Bierstedt, B. Kramm, R. Meise, M. Schottenloher and D. Vogt, the Italian school as represented by E. Vesentini, and the Swedish school, as represented by C.O. Kiselman.

Let us speak a little about the contents of this book. We begin with a study of algebraic and topological properties of m-linear mappings, m-homogeneous polynomials and power series, and then introduce the concept of holomorphic mapping between complex normed spaces and complex locally convex spaces. We endorse Weierstrass's point of view, that is, that holomorphic mappings are, in a sense, locally represented by their Taylor series. Several expressions are then derived for Cauchy's integral formula and for Cauchy's inequalities; then a study is presented of the convergence of Taylor series of a holomorphic mapping. The differences with the case of infinite dimension are stressed, thus leading naturally to a consideration of holomorphic mappings of bounded type. The relationships are shown between the notions of holomorphic mapping, weakly holomorphic mapping and finitely holomorphic mapping (holomorphic in Gâteaux's sense). We then present the infinite-dimensional versions of the theorem of maximum module and uniqueness of holomorphic continuation. In studying the $\mathcal{H}$-bounding sets of locally convex spaces, we apply the Josefson-Nissenzweig theorem: "If E is a normed space of infinite dimension, there exists a sequence $\{\varphi_m\}_{m\in\mathbb{N}}$ in E' such that $\|\varphi_m\| = 1$ for every $m \in \mathbb{N}$ and $\varphi_m \to 0$ as $m \to \infty$ in the weak topology $\sigma(E',E)$". This theorem resolves a famous problem proposed by Banach, and thus enjoys an application in an area different from its initial context. A detailed study is made of the properties of topologies τ_o, τ_ω, τ_δ both in Part I and Part II. The spaces of the bounded type holomorphic mappings are dealt with in detail and in the context of such spaces we prove the Cartan-Thullen theory in the case of the base space being separable, this bringing Part I to a close. Part II ends with the study of bornological properties of the spaces of holomorphic mappings.

Incomparably more could be said about the topics that have been left out. Particular mention should be made of the fact that no property of a nuclear nature is referred to in the text, although questions relating to nuclearity are becoming more and more important in the study of Holomorphy.

Profitable use of this book will require some familarity with the basic theorems of Functional Analysis, in the context of normed spaces and locally convex spaces. The reading of Part II does not presume knowledge of Part I. Theoretically, it could be said that the study of both parts requires no previous knowledge of several complex variables. This, however, in the author's opinion, is a marvellous case of wishful thinking.

This work owes much to the experience acquired during the courses administered at the Federal University of Rio de Janeiro, at the University of Santiago de Compostela (on invitation of Professor J.M. Isidro) and at the University of Valencia (on invitation of Professor M. Valdivia). But it owes most to what was learned in the classes given by Professor Leopoldo Nachbin, to whom we are deeply grateful.

A special word of thanks to Dr. Raymond Ryan of the University of Galway, Ireland, for his English translation of a preliminary version of this book. Many thanks also to Professors S. Dineen and J.P. Ansemil for their suggestions. Our thanks also go to the Mathematics Institute of the Federal University of Rio de Janeiro for their financial support, and to Wilson Góes for his excellent typing services.

Jorge Alberto Barroso

Federal University of Rio de Janeiro
August 1984

TABLE OF CONTENTS

FOREWORD vii

PART I. THE NORMED CASE

CHAPTER 1. NOTATION AND TERMINOLOGY. POLYNOMIALS 1
CHAPTER 2. POWER SERIES 17
CHAPTER 3. HOLOMORPHIC MAPPINGS 25
CHAPTER 4. THE CAUCHY INTEGRAL FORMULAS 31
CHAPTER 5. CONVERGENCE OF THE TAYLOR SERIES 45
CHAPTER 6. WEAK HOLOMORPHY 57
CHAPTER 7. FINITE HOLOMORPHY AND GATEAUX HOLOMORPHY ... 69
CHAPTER 8. TOPOLOGIES ON SPACES OF HOLOMORPHIC MAPPINGS 81
CHAPTER 9. UNIQUENESS OF ANALYTIC CONTINUATION 111
CHAPTER 10. THE MAXIMUM PRINCIPLE 115
CHAPTER 11. HOLOMORPHIC MAPPINGS OF BOUNDED TYPE 119
CHAPTER 12. DOMAINS OF $\mathcal{H}_b$-HOLOMORPHY 127
CHAPTER 13. THE CARTAN-THULLEN THEOREM FOR DOMAINS OF $\mathcal{H}_b$-HOLOMORPHY 139

PART II. THE LOCALLY CONVEX CASE

CHAPTER 14. NOTATION AND MULTILINEAR MAPPINGS............ 155
CHAPTER 15. POLYNOMIALS 159
CHAPTER 16. TOPOLOGIES ON SPACES OF MULTILINEAR MAPPINGS AND HOMOGENEOUS POLYNOMIALS 167
CHAPTER 17. FORMAL POWER SERIES 173
CHAPTER 18. HOLOMORPHIC MAPPINGS 177
CHAPTER 19. SEPARATION AND PASSAGE TO THE QUOTIENT 183
CHAPTER 20. $\mathcal{H}$-HOLOMORPHY AND H-HOLOMORPHY 185
CHAPTER 21. ENTIRE MAPPINGS 187
CHAPTER 22. SOME ELEMENTARY PROPERTIES OF HOLOMORPHIC MAPPINGS 191

CHAPTER 23. HOLOMORPHY, CONTINUITY AND AMPLE BOUNDEDNESS 195

CHAPTER 24. BOUNDING SETS 197

CHAPTER 25. THE CAUCHY INTEGRAL AND THE CAUCHY INEQUALITIES 209

CHAPTER 26. THE TAYLOR REMAINDER 215

CHAPTER 27. COMPACT AND LOCAL CONVERGENCE OF THE TAYLOR SERIES 221

CHAPTER 28. THE MULTIPLE CAUCHY INTEGRAL AND THE CAUCHY INEQUALITIES 229

CHAPTER 29. DIFFERENTIALLY STABLE SPACES 233

CHAPTER 30. LIMITS OF HOLOMORPHIC MAPPINGS 237

CHAPTER 31. UNIQUENESS OF HOLOMORPHIC CONTINUATION 241

CHAPTER 32. HOLOMORPHY AND FINITE HOLOMORPHY 245

CHAPTER 33. THE MAXIMUM SEMINORM THEOREM 249

CHAPTER 34. PROJECTIVE AND INDUCTIVE LIMITS AND HOLOMORPHY 253

CHAPTER 35. TOPOLOGIES ON $\mathcal{H}(U;F)$ 263

CHAPTER 36. BOUNDED SUBSETS OF $\mathcal{H}(U;F)$ 273

BIBLIOGRAPHY 279

AN INDEX OF DEFINITIONS 297

AUTHOR INDEX 301

PART I

THE NORMED CASE

CHAPTER 1

NOTATION AND TERMINOLOGY. POLYNOMIALS

We denote by $\mathbb{N}$, $\mathbb{R}$ and $\mathbb{C}$ the systems of non-negative integers, real numbers and complex numbers respectively. Throughout this book, all vector spaces considered will have $\mathbb{C}$ as their field of scalars unless explicitly stated otherwise. E and F will denote complex normed spaces, and U a non-empty open subset of E.

If ξ is a point in a normed space and ρ a positive real number, the open ball (respectively, the closed ball) in this space with centre ξ and radius ρ will be denoted by $B_\rho(\xi)$ (respectively $\bar{B}_\rho(\xi)$).

DEFINITION 1.1 Let $E_1, E_2, \ldots, E_m$ ($m \in \mathbb{N}$, $m > 0$) be a finite sequence of normed spaces.

$\mathcal{L}_a(E_1, \ldots, E_m; F)$ denotes the vector space of m-linear mappings of $\prod_{i=1}^{m} E_i = E_1 \times \ldots \times E_m$ into F, where addition and multiplication by scalars are defined pointwise.

$\mathcal{L}(E_1, \ldots, E_m; F)$ denotes the subspace of $\mathcal{L}_a(E_1, \ldots, E_m; F)$ of continuous m-linear mappings of $\prod_{i=1}^{m} E_i$ into F. ($\prod_{i=1}^{m} E_i$ being endowed with the product topology). In the case $m = 0$ we identify $\mathcal{L}_a(E_1, \ldots, E_m; F)$ with F as vector spaces, and $\mathcal{L}(E_1, \ldots, E_m; F)$ with F as normed spaces.

Let $A \in \mathcal{L}(E_1,\ldots,E_m;F)$. It is easy to see that

1)
$$\sup_{x_1\neq 0,\ldots,x_m\neq 0} \frac{\|A(x_1,\ldots,x_m)\|}{\|x_1\|\ldots\|x_m\|} =$$

$$\sup_{\|x_1\|\leq 1,\ldots,\|x_m\|\leq 1} \|A(x_1,\ldots,x_m)\| =$$

$\inf\{M \geq 0\colon \|A(x_1,\ldots,x_m)\| \leq M\|x_1\|\ldots\|x_m\|$ for all $x_1\in E_1,\ldots,x_m\in E_m\}$,

and

2) denoting by $\|A\|$ the common value of the expressions which appear in 1), the mapping

$$A \in \mathcal{L}(E_1,\ldots,E_m;F) \mapsto \|A\|$$

is a norm.

REMARK 1.1: a) We commit an abuse of notation, using the same symbol to represent the norms on spaces which may be distinct.

b) If $A \in \mathcal{L}(E_1,\ldots,E_m;F)$, it is easy to see that $\|A(x_1,\ldots,x_m)\| \leq \|A\|\|x_1\|\ldots\|x_m\|$ for every $x_1\in E_1,\ldots,x_m\in E_m$.

c) If $A \in \mathcal{L}_a(E_1,\ldots,E_m;F)$ then $A \in \mathcal{L}(E_1,\ldots,E_m;F)$ if and only if

$$\sup_{x_1\neq 0,\ldots,x_m\neq 0} \frac{\|A(x_1,\ldots,x_m)\|}{\|x_1\|\ldots\|x_m\|} < \infty.$$

d) In the case in which F is a Banach space, so too is the space $\mathcal{L}(E_1,\ldots,E_m;F)$, with the norm above.

In the case in which $E_1 = E_2 = \ldots = E_m = E$ the space of m-linear mappings (respectively, of continuous m-linear mappings) of $E^m = \overbrace{E\times\ldots\times E}^{m}$ into F will be denoted by $\mathcal{L}_a({}^mE;F)$ (respectively, $\mathcal{L}({}^mE;F)$).

We denote by $\mathcal{L}_{as}({}^mE;F)$ $(m > 0)$ the subspace of $\mathcal{L}_a({}^mE;F)$ consisting of the symmetric m-linear mappings of E^m into F. In other words, if S_m is the symmetric group of order m, then

$$\mathcal{L}_{as}({}^mE;F) = \{A \in \mathcal{L}_a({}^mE;F) : A(x_1,\ldots,x_m) = A(x_{\sigma(1)},\ldots,x_{\sigma(m)})$$
$$\text{for every } x_1,\ldots,x_m \in E \text{ and } \sigma \in S_m\}.$$

In the case $m = 0$, $\mathcal{L}_{as}({}^mE;F) = \mathcal{L}_a({}^mE;F) = F$ as vector spaces and $\mathcal{L}_s({}^mE;F) = \mathcal{L}^m(E;F) = F$ as normed spaces.

We denote by $\mathcal{L}_s({}^mE;F)$ the subspace of $\mathcal{L}({}^mE;F)$ consisting of the continuous symmetric m-linear mappings of E^m into F. Thus,

$$\mathcal{L}_s({}^mE;F) = \mathcal{L}({}^mE;F) \cap \mathcal{L}_{as}({}^mE;F).$$

It is easy to see that $\mathcal{L}_s({}^mE;F)$ is a closed subspace of $\mathcal{L}({}^mE;F)$.

With each $A \in \mathcal{L}_a({}^mE;F)$ is associated an element of $\mathcal{L}_{as}({}^mE;F)$ which we denote by A_s, and call the symmetrization of A, defined by:

$$A_s(x_1,\ldots,x_m) = \frac{1}{m!}\sum_{\sigma\in S_m} A(x_{\sigma(1)},\ldots,x_{\sigma(m)})$$

for $x_1,\ldots,x_m \in E$. The mapping $A \in \mathcal{L}_a({}^mE;F) \mapsto A_s \in \mathcal{L}_{as}({}^mE;F)$ is linear and surjective, and is a projection of $\mathcal{L}_a({}^mE;F)$ onto the subspace $\mathcal{L}_{as}({}^mE;F)$; thus $(A_s)_s = A_s$ for every $A \in \mathcal{L}_a({}^mE;F)$.

By restriction we obtain a projection of $\mathcal{L}(^mE;F)$ onto $\mathcal{L}_s(^mE;F)$ which is continuous since $\|A_s\| \leq \|A\|$ for every $A \in \mathcal{L}(^mE;F)$.

If $m \in \mathbb{N}$, $m > 0$, and $A \in \mathcal{L}_a(^mE;F)$, we denote $A(\overbrace{x,\dots,x}^{m})$ by Ax^m; if $m = 0$ we write $Ax^0 = A$ for every $x \in E$.

DEFINITION 1.2 Let $m \in \mathbb{N}$. A mapping $P: E \to F$ is called an m-homogeneous polynomial from E into F if there exists $A \in \mathcal{L}_a(^mE;F)$ such that $P(x) = Ax^m$ for every $x \in E$. When P and A are related in this way we write $P = \hat{A}$.

REMARK 1.2: If $A \in \mathcal{L}_a(^mE;F)$ and $P = \hat{A}$, then P is the restriction of A to the diagonal of E^m, since $P(x) = A(\overbrace{x,\dots,x}^{m})$. If $\Delta_m: E \to E^m$ is the diagonal mapping, $\Delta_m(x) = (\overbrace{x,\dots,x}^{m})$ for every $x \in E$, then $P = \hat{A}$ is equivalent to $P = A \circ \Delta_m$.

We denote by $\mathcal{P}_a(^mE;F)$ the set of m-homogeneous polynomials from E into F. This set forms a vector space, addition and multiplication by scalars being defined pointwise.

DEFINITION 1.3 Let $m \in \mathbb{N}$. A mapping $P: E \to F$ is called a continuous m-homogeneous polynomial from E into F if there exists $A \in \mathcal{L}(^mE;F)$ such that $P(x) = Ax^m$ for every $x \in E$.

We denote by $\mathcal{P}(^mE;F)$ the set of continuous m-homogeneous polynomials from E into F. This set forms a vector space, addition and multiplication by scalars being defined pointwise, and is a subspace of $\mathcal{P}_a(^mE;F)$.

It is easy to see that if $P \in \mathcal{P}(^mE;F)$ then

$$\sup_{x\in E, x\neq 0} \frac{\|P(x)\|}{\|x\|^m} = \sup_{x\in E, \|x\|\leq 1} \|P(x)\|$$

$$= \inf\{M \geq 0 : \|P(x)\| \leq M\|x\|^m \text{ for every } x \in E\} < \infty,$$

and that the mapping $P \in \mathcal{P}(^mE;F) \mapsto \|P\|$ is a norm, where $\|P\|$ denotes the common value of the expressions above.

REMARK 1.3: a) In the case in which F is a Banach space, so too is the space $\mathcal{P}(^mE;F)$ with the norm above.

b) If $P \in \mathcal{P}(^mE;F)$, then $\|P(x)\| \leq \|P\|\|x\|^m$ for every $x \in E$.

c) In the case $m = 0$, $\mathcal{P}_a(^0E;F)$ is identified with F as a vector space, and $\mathcal{P}(^mE;F)$ is identified with F as a normed space.

EXAMPLE 1.1 In the case in which $E = F = \mathbb{C}$, every m-homogeneous polynomial from E into F is of the form $P(\lambda) = a\lambda^m$ for every $\lambda \in \mathbb{C}$, where a is some element of $\mathbb{C}$. More generally, taking $E = \mathbb{C}$ and F an arbitrary normed space, then every m-homogeneous polynomial is of the form $P(\lambda) = b\lambda^m$ for every $\lambda \in \mathbb{C}$, where b is some element of F.

In fact, every mapping $A \in \mathcal{L}_a(^m\mathbb{C};F)$ is of the form $A(\lambda_1,\ldots,\lambda_m) = \lambda_1 \ldots \lambda_m b$ for all $\lambda_1,\ldots,\lambda_m \in \mathbb{C}$, where b is some element of F, and thus $P = \hat{A}$ takes the form $P(\lambda) = b\lambda^m$, $\lambda \in \mathbb{C}$.

REMARK 1.4: If $A \in \mathcal{L}_a(^mE;F)$ and A_s is its symmetrization, then $\hat{A} = \hat{A}_s$. To see this, let $x_1,\ldots,x_m \in E$. Then

$$A_s(x_1,\ldots,x_m) = \frac{1}{m!} \sum_{\sigma \in S_m} A(x_{\sigma(1)},\ldots,x_{\sigma(m)}),$$

and taking $x_1 = x_2 = \ldots = x_m = x$ yields

$$\hat{A}_s(x) = A_s(\overbrace{x,\ldots,x}^{m}) = \frac{1}{m!} \sum_{\sigma \in S_m} A(x_{\sigma(1)},\ldots,x_{\sigma(m)})$$

$$= \frac{1}{m!}\, m!A(x,\ldots,x) = A(\overbrace{x,\ldots,x}^{m}) = \hat{A}(x).$$

PROPOSITION 1.1 (The Polarization Formula):

Let $m \in \mathbb{N}$, $m \geq 1$, $A \in \mathcal{L}_{as}(^mE;F)$ and $P = \hat{A}$. Then for $x_1,\ldots,x_m \in E$,

$$A(x_1,\ldots,x_m) = \frac{1}{2^m m!} \sum_{\substack{\epsilon_i = \pm 1 \\ 1 \leq i \leq m}} \epsilon_1 \cdots \epsilon_m \, P(\epsilon_1 x_1 + \ldots + \epsilon_m x_m).$$

We omit the proof of this formula which is purely algebraic in nature.

PROPOSITION 1.2: a) The mapping

$$A \in \mathcal{L}_a(^mE;F) \mapsto \hat{A} \in \mathcal{P}_a(^mE;F)$$

is linear and surjective for every $m \in \mathbb{N}$.

b) The mapping

$$A \in \mathcal{L}_{as}(^mE;F) \mapsto \hat{A} \in \mathcal{P}_a(^mE;F)$$

is an isomorphism of vector spaces for every $m \in \mathbb{N}$.

PROOF: a) Surjectivity of this mapping is an immediate consequence of the definition of an m-homogeneous polynomial;

the proof of linearity is trivial.

b) This mapping is certainly linear, being the restriction of the mapping given in a) to the subspace $\mathcal{L}_{as}({}^mE;F)$ of $\mathcal{L}_a({}^mE;F)$.

The mapping is surjective since, given $P \in \mathcal{P}_a({}^mE;F)$, by part a) there exists $A \in \mathcal{L}_a({}^mE;F)$ such that $P = \hat{A}$, and by Remark 1.4, $\hat{A}_s = \hat{A} = P$, and $A_s \in \mathcal{L}_{as}({}^mE;F)$.

To see that this mapping is injective, let $A \in \mathcal{L}_{as}({}^mE;F)$.

By the polarization formula,

$$A(x_1,\ldots,x_m) = \frac{1}{2^m m!} \sum_{\substack{\varepsilon_i=\pm 1 \\ 1\leq i\leq m}} \varepsilon_1\ldots\varepsilon_m \hat{A}(\varepsilon_1 x_1+\ldots+\varepsilon_m x_m)$$

for all $x_1,\ldots,x_m \in E$. Thus if $\hat{A} = 0$, it follows that $A(x_1,\ldots,x_m) = 0$ for all $x_1,\ldots,x_m \in E$, and hence $A = 0$.

Q.E.D.

PROPOSITION 1.3: a) The mapping

$$A \in \mathcal{L}({}^mE;F) \mapsto \hat{A} \in \mathcal{P}({}^mE;F)$$

is linear, surjective and continuous.

b) The mapping

$$A \in \mathcal{L}_s({}^mE;F) \mapsto \hat{A} \in \mathcal{P}({}^mE;F)$$

is an isomorphism of vector spaces and a homeomorphism. Furthermore,

$$\|\hat{A}\| \leq \|A\| \leq \frac{m^m}{m!}\|\hat{A}\|$$

for every $A \in \mathcal{L}_s(^mE;F)$ and $m \in \mathbb{N}$.

PROOF: a) This mapping takes its values in $\mathcal{P}(^mE;F)$ and is surjective by the definition of a continuous m-homogeneous polynomial; it is linear since it is the restriction to the subspace $\mathcal{L}(^mE;F)$ of $\mathcal{L}_a(^mE;F)$ of a linear mapping defined on $\mathcal{L}_a(^mE;F)$. Continuity is a consequence of the inequality $\|\hat{A}\| \leq \|A\|$, whose verification is immediate.

b) This mapping is certainly linear, being the restriction of the mapping given in a) to the subspace $\mathcal{L}_s(^mE;F)$ of $\mathcal{L}(^mE;F)$, and is continuous for the same reason.

The mapping is surjective since, given $P \in \mathcal{P}(^mE;F)$, by part a) there exists $A \in \mathcal{L}(^mE;F)$ such that $P = \hat{A}$. We know that $\hat{A}_s = \hat{A}$ (Remark 1.4), and it is easy to see that if $A \in \mathcal{L}(^mE;F)$, then $A_s \in \mathcal{L}_s(^mE;F)$. Injectivity is a consequence of part b) of Proposition 1.2, since this mapping is the restriction of the mapping considered there to the subspace $\mathcal{L}_s(^mE;F)$ of $\mathcal{L}_{as}(^mE;F)$.

Finally, we prove the inequality $\|\hat{A}\| \leq \|A\| \leq \frac{m^m}{m!}\|\hat{A}\|$, from which it follows that the given mapping is a homeomorphism. We show that $\|A\| \leq \frac{m^m}{m!}\|\hat{A}\|$ - as we have already indicated, the other inequality is immediate. Let $A \in \mathcal{L}_s(^mE;F)$ and $x_1,\ldots,x_m \in E$. By the polarization formula,

$$A(x_1,\ldots,x_m) = \frac{1}{2^m m!} \sum_{\substack{\varepsilon_i=\pm 1\\ 1\leq i\leq m}} \varepsilon_1\cdots\varepsilon_m\, \hat{A}(\varepsilon_1 x_1+\cdots+\varepsilon_m x_m),$$

whence

$$\|A(x_1,\ldots,x_m)\| \le \frac{1}{2^m m!} \sum_{\substack{\varepsilon_i=\pm 1\\ 1\le i\le m}} \|\hat{A}(\varepsilon_1 x_1+\ldots+\varepsilon_m x_m)\|$$

$$\le \frac{1}{2^m m!} \sum_{\substack{\varepsilon_i=\pm 1\\ 1\le i\le m}} \|\hat{A}\| \|\varepsilon_1 x_1+\ldots+\varepsilon_m x_m\|^m$$

$$\le \frac{1}{2^m m!} \sum_{\substack{\varepsilon_i=\pm 1\\ 1\le i\le m}} \|\hat{A}\| (\|x_1\|+\ldots+\|x_m\|)^m.$$

Thus, if $\|x_1\| = \ldots = \|x_m\| = 1$, then

$$\|A(x_1,\ldots,x_m)\| \le \frac{1}{2^m m!} \sum_{\substack{\varepsilon_i=\pm 1\\ 1\le i\le m}} \|\hat{A}\| m^m = \frac{1}{2^m m!} 2^m \|\hat{A}\| m^m$$

$$= \frac{m^m}{m!} \|\hat{A}\|,$$

and hence

$$\|A\| \le \frac{m^m}{m!} \|\hat{A}\|.$$

Q.E.D.

REMARK 1.5: The mapping $A \in \mathcal{L}_s(^mE;F) \mapsto \hat{A} \in \mathcal{P}(^mE;F)$ is not in general an isometry. In fact the smallest constant C which satisfies: $\|A\| \le C\|\hat{A}\|$ independently of E and F, depending only on m, is $\frac{m^m}{m!}$. This is shown by the following example.

EXAMPLE 1.2 Let E be ℓ^1, the vector space of all sequences $x = (x_1,x_2,\ldots,x_m,\ldots)$ of complex numbers for which $\sum_{n=1}^{\infty} |x_n| < \infty$, with the norm $\|x\| = \sum_{n=1}^{\infty} |x_n|$, and let $F = \mathbb{C}$.

Let m be a positive integer, $m \ge 1$, and let $x^1 = (x_1^1,x_2^1,\ldots,x_m^1,\ldots)$, $x^2 = (x_1^2,x_2^2,\ldots,x_m^2,\ldots),\ldots,x^m = (x_1^m,x_2^m,\ldots,x_m^m,\ldots)$ be elements of E. We define $A_m:E^m \to \mathbb{C}$

by:

$$A_m(x^1,x^2,\dots,x^m) = \frac{1}{m!} \sum_{\sigma\in S_m} x_1^{\sigma(1)} x_2^{\sigma(2)} \dots x_m^{\sigma(m)} .$$

It is easy to see that A_m is a symmetric m-linear mapping from E^m into $\mathbb{C}$. We have

$$\|A_m(x^1,x^2,\dots,x^m)\| \le \frac{1}{m!} \sum_{\sigma\in S_m} |x_1^{\sigma(1)}||x_2^{\sigma(2)}|\dots|x_m^{\sigma(m)}|$$

$$\le \frac{1}{m!} \left(\sum_{j=1}^{m} |x_j^1|\right)\left(\sum_{j=1}^{m} |x_j^2|\right)\dots\left(\sum_{j=1}^{m} |x_j^m|\right)$$

$$\le \frac{1}{m!} \|x^1\|\|x^2\|\dots\|x^m\| .$$

Thus

$$(1) \qquad \|A_m\| = \sup_{\|x^1\|\le\dots\le\|^m\| 1} |A_m(x^1,x^2,\dots,x^m)| \le \frac{1}{m!} .$$

Furthermore, taking $x^1 = (1,0,\dots,0,\dots)$, $x^2 = (0,1,0,\dots,0,\dots)$, $,\dots,x^m = (0,\dots,0,\overset{(m)}{1},0,\dots)$, we have $x^1,x^2,\dots,x^m \in E$ and

$$(2) \qquad \|A_m(x^1,x^2,\dots,x^m)\| = \frac{1}{m!} .$$

(1) and (2) together imply that $\|A_m\| = \frac{1}{m!}$.

Now let $x \in E$, $x = (x_1,x_2,\dots,x_m,\dots)$. Then $\hat{A}_m(x) = A_m(\overbrace{x,\dots,x}^{m}) = x_1x_2\dots x_m$. Thus for $x \in E$ we have

$$|\hat{A}_m(x)| = |x_1||x_2|\dots|x_m| \le \frac{(|x_1|+\dots+|x_m|)^m}{m^m} ,$$

which implies that

$$\|\hat{A}\| = \sup_{x\neq 0} \frac{|\hat{A}_m(x)|}{\|x\|^m} \le \sup_{x\neq 0} \frac{(|x_1|+\dots+|x_m|)^m}{\|x\|^m\, m^m} \le \frac{1}{m^m} .$$

Taking $x = (\overbrace{\frac{1}{m},\frac{1}{m},\ldots,\frac{1}{m}}^{m},0,\ldots,0,\ldots) \in E$, we have $\|x\| = 1$ and $|\hat{A}_m(x)| = \frac{1}{m^m}$. Therefore $\|\hat{A}_m\| = \frac{1}{m^m}$.

Thus we have shown that $\|A_m\| = \frac{m^m}{m!}\|\hat{A}_m\|$.

If E and F are vector spaces we shall denote by $\mathfrak{F}(E;F)$ the set of all mappings from E into F. This set forms a vector space, addition and multiplication by scalars being defined pointwise.

REMARK 1.6: Let B be a vector space and $(B_n)_{n\in\mathbb{N}}$ a sequence of subspaces of B. We denote by $\sum_{m\in\mathbb{N}} B_m$ the set of the finite sums which can be formed with elements chosen from the subspaces B_m. In other words, $x \in \sum_{m\in\mathbb{N}} B_m$ if there exist integers $m_1,m_2,\ldots,m_k$ and elements $x_{m_1} \in B_{m_1}$, $x_{m_2} \in B_{m_2},\ldots,x_{m_k} \in B_{m_k}$ such that $x = x_{m_1} + x_{m_2} + \ldots + x_{m_k}$. Then $\sum_{m\in\mathbb{N}} B_m$ is a subspace of B - it is referred to as the algebraic sum of the subspaces B_m.

An algebraic sum of subspaces B_m, $m \in \mathbb{N}$, is called a direct algebraic sum if, whenever the m_j are pairwise distinct, the condition $x = 0$ implies $x_{m_1} = x_{m_2} = \ldots = x_{m_k} = 0$. In the case of a direct algebraic sum, we employ the notation $\bigoplus_{m\in\mathbb{N}} B_m$, and we say that the subspaces B_m of B, $m \in \mathbb{N}$, are linearly independent.

Now let E and F be normed spaces, and let us consider the sequences $\mathcal{P}_a(^mE;F)$ and $\mathcal{P}(^mE;F)$, $m \in \mathbb{N}$, and their algebraic sums $\sum_{m\in\mathbb{N}} \mathcal{P}_a(^mE;F)$ and $\sum_{m\in\mathbb{N}} \mathcal{P}(^mE;F)$ within the vector space $\mathcal{F}(E;F)$.

PROPOSITION 1.4. The subspaces $\mathcal{P}_a(^mE;F)$, $m \in \mathbb{N}$, of $\mathcal{F}(E;F)$ are linearly independent.

PROOF: It suffices to prove the following statement for every $m \in \mathbb{N}$: if $P_j \in \mathcal{P}_a(^jE;F)$, $j = 0,1,\ldots,m$, and $P = P_o + P_1 + \ldots + P_m$, the condition $P = 0$ implies $P_j = 0$ for $j = 0,1,\ldots,m$. We shall prove this by induction. For $m = 0$ there is nothing to prove. We assume the truth of the statement for $m-1$, $m \geq 1$, and prove it for m. Accordingly, let $P_j \in \mathcal{P}(^jE;F)$, $j = 0,1,\ldots,m$, and suppose that $\sum_{j=0}^{m} P_j = 0$. Then for all $x \in E$ and $\lambda \in \mathbb{C}$ we have

$$\text{(a)} \quad \sum_{j=0}^{m} P_j(x) = 0 \qquad \text{and} \qquad \text{(b)} \quad \sum_{j=0}^{m} P_j(\lambda x) = 0.$$

Multiplying equation (a) by λ^m and subtracting the result from equation (b) we obtain:

$$\text{(c)} \quad \sum_{j=0}^{m-1} (\lambda^m-\lambda^j)P_j(x) = 0 \quad \text{for all } \lambda \in \mathbb{C} \text{ and } x \in E.$$

We now choose a value of $\lambda \in \mathbb{C}$ which is not a solution of any of the m equations $\lambda^m-\lambda^j = 0$, $j = 0,1,\ldots,m-1$. Then the induction hypothesis applied to the relation $\sum_{j=0}^{m-1} (\lambda^m-\lambda^j)P_j = 0$ given in (c) shows that $P_j = 0$ for $j = 0,1,\ldots,m-1$, and then, since $\sum_{j=0}^{m} P_j = 0$, we also have $P_m = 0$. Q.E.D.

COROLLARY 1.1 The subspaces $\mathcal{P}(^mE;F)$, $m \in \mathbb{N}$, of $\mathcal{F}(E;F)$ are linearly independent.

PROOF: This is a straightforward application of the proposition, using the fact that $\mathcal{P}(^mE;F) \subset \mathcal{P}_a(^mE;F)$ for every $m \in \mathbb{N}$.

Q.E.D.

REMARK 1.7 In the language of algebraic sums and direct algebraic sums Proposition 1.4 and its Corollary state that

$$\sum_{m \in \mathbb{N}} \mathcal{P}_a(^mE;F) = \bigoplus_{m \in \mathbb{N}} \mathcal{P}_a(^mE;F)$$

and

$$\sum_{m \in \mathbb{N}} \mathcal{P}(^mE;F) = \bigoplus_{m \in \mathbb{N}} \mathcal{P}(^mE;F).$$

We denote by $\mathcal{P}_a(E;F)$ (respectively, $\mathcal{P}(E;F)$) the subspace $\bigoplus_{m \in \mathbb{N}} \mathcal{P}_a(^mE;F)$ (respectively, $\bigoplus_{m \in \mathbb{N}} \mathcal{P}(^mE;F)$).

DEFINITION 1.4 An element of $\mathcal{P}_a(E;F)$ is called a polynomial from E into F, and an element of $\mathcal{P}(E;F)$ is called a continuous polynomial from E into F.

Thus to say that P is a polynomial from E into F means that either $P = 0$, or if $P \neq 0$, that P can be written in a unique way in the form $P = P_0 + P_1 + \ldots + P_m$, where $P_j \in \mathcal{P}_a(^jE;F)$, $j = 0,1,\ldots,m$, and $P_m \neq 0$. In the latter case the natural number m is called the degree of the polynomial P. By convention the polynomial $P = 0$ is assigned the degree -1:

PROPOSITION 1.5 If $P \in \mathcal{P}_a(E;F)$ and $P = \sum_{j=0}^{m} P_j$, where $P_j \in \mathcal{P}_a(^jE;F)$, then $P \in \mathcal{P}(E;F)$ if and only if $P_j \in \mathcal{P}(^jE;F)$ for every $j = 0,1,\ldots,m$.

The proof of this proposition is similar to that of Proposition 1.4 and will be left to the reader to carry out, as will the proof of the following:

PROPOSITION 1.6 Let E, F and G be normed spaces.

(a) If $P \in \mathcal{P}_a(E;F)$ and $Q \in \mathcal{P}_a(F;G)$, then $Q \circ P \in \mathcal{P}_a(E;G)$.

(b) If $P \in \mathcal{P}(E;F)$ and $Q \in \mathcal{P}(F;G)$, then $Q \circ P \in \mathcal{P}(E;G)$.

We state without proof the following:

"PROPOSITION A: If $A \in \mathcal{L}_s(^mE;F)$, $m \in \mathbb{N}$, the following are equivalent:

a) A is continuous.

b) A is continuous at one point of E^m.

c) There exists a neighborhood V of the origin in E^m such that A is bounded in V."

This is used to prove

PROPOSITION 1.7 If $P \in \mathcal{P}_a(E;F)$ the following are equivalent:

a) P is continuous.

b) P is continuous at one point of E.

c) There exists a non-empty open subset U of E such that P is bounded in U.

PROOF: The implications a) $\Rightarrow$ b) $\Rightarrow$ c) are easily verified. We prove the implication c) $\Rightarrow$ a).

By c), there exists a non-empty open subset U of E and a constant $M \geq 0$ such that

1) $\|P(x)\| \leq M$ for every $x \in U$.

Since U is non-empty there exists $x_o \in U$. Let $t: E \to E$ be the translation: $t(x) = x - x_o$, $x \in E$. Then 1) is equivalent to

2) $\|P\circ t^{-1}(y)\| \leq M$ for every $y \in V = t(U)$ and V is a neighborhood of the origin in E. Since t is a continuous polynomial from E into E (of degree 1) we have, by Proposition 1.6, that $Q = P\circ t^{-1}$ is a polynomial, and, by 2), $\|Q(y)\| \leq M$ for every $y \in V$.

As $Q = P\circ t^{-1}$ and $P = Q\circ t$, continuity of P is equivalent to continuity of Q. We shall prove that Q is continuous. We begin with the following assertion:

(*) "If $Q = \sum_{j=0}^{m} Q_j$, where $Q_j \in \mathcal{P}_a({}^jE;F)$ for $j = 0,1,\ldots,m$, and $Q_m \neq 0$, then Q is bounded in a subset V of E if and only if Q_j is bounded in V for every j, $j = 0,1,\ldots,m$."

If the Q_j, $j = 0,1,\ldots,m$, are bounded in V then Q is certainly bounded in V also. We prove the converse by induction. The case $m = 0$ is trivial. Assuming the truth of the statement for all natural numbers less than m, $m \geq 1$, we consider the case m. Then:

3) $Q(x) = \sum_{j=0}^{m} Q_j(x)$ and

4) $Q(\lambda x) = \sum_{j=0}^{m} Q_j(\lambda x)$

for all $x \in E$ and $\lambda \in C$. From 3) and 4) we obtain

5) $\lambda^m Q(x) - Q(\lambda x) = \sum_{j=0}^{m-1} (\lambda^m - \lambda^j) Q_j(x)$

for all $x \in E$ and $\lambda \in C$. We choose a value for $\lambda \in \mathbb{C}$ which is not a solution of any of the equations $\lambda^m - \lambda^j = 0$,

$j = 0,1,\ldots,m-1$. The polynomial

$$\sum_{j=0}^{m-1} (\lambda^m - \lambda^j) Q_j(x)$$

has degree at most $m-1$, and is bounded in V since, by hypothesis, Q is bounded in V, and hence $\lambda^m Q(x) - Q(\lambda x)$ is bounded in V for each fixed value of λ. With λ chosen as indicated, it now follows from the induction hypothesis that Q_j is bounded in V for $j = 0,1,\ldots,m-1$, and then since

$$6) \quad Q_m = Q - \sum_{j=0}^{m-1} Q_j ,$$

Q_m is also bounded in V. This proves assertion (*).

Let $A_j \in \mathcal{L}_{as}({}^jE;F)$ with $Q_j = \hat{A}_j$ for $j = 0,1,\ldots,m$. Then

$$7) \quad Q(x) = \sum_{j=0}^{m} Q_j(x) = \sum_{j=0}^{m} A_j x^j .$$

Now let W be a balanced neighborhood of the origin in E such that

$$\overbrace{W + \ldots + W}^{j} \subset V$$

for every $j = 1,\ldots,m$. Then if $(x_1,\ldots,x_j) \in W^j$ we have $\sum_{i=1}^{j} \varepsilon_i x_i \in V$ if $\varepsilon_i = \pm 1$, for every $j = 1,\ldots,m$. Using the polarization formula, we conclude that A_j is bounded in the neighborhood W^j of the origin in E^j, for $j = 1,\ldots,m$. A_o is constant, and hence is continuous.

It now follows from Proposition A above that A_j is continuous for every $j = 0,1,\ldots,m$. Therefore, by relation 7), Q is continuous. Q.E.D.

CHAPTER 2

POWER SERIES

DEFINITION 2.1 A power series from E into F about the point $\xi \in E$ is a series in the variable $x \in E$ of the form

$$\sum_{m=0}^{\infty} A_m(x-\xi)^m$$

where $A_m \in \mathcal{L}_s({}^mE;F)$ for $m = 0,1,\ldots$; if we prefer to use polynomials the series can be written:

$$\sum_{m=0}^{\infty} P_m(x-\xi)$$

where $P_m = \hat{A}_m$ for $m = 0,1,\ldots$. The A_m, or the corresponding polynomials P_m, are often referred to as the coefficients of the power series in question.

DEFINITION 2.2 The radius of convergence, or the radius of uniform convergence, of a power series about the point $\xi \in E$ is the supremum of the set of numbers r, $0 \leq r \leq \infty$ such that the series is uniformly convergent in $\bar{B}_\rho(\xi)$ for every ρ, $0 \leq \rho < r$.

A power series is said to be convergent, or uniformly convergent, when its radius of convergence is greater than zero, that is, when there exists $\rho > 0$ such that the series converges uniformly in $\bar{B}_\rho(\xi)$.

For power series from $\mathbb{C}$ into $\mathbb{C}$, the radius of convergence can be calculated by means of the Cauchy-Hadamard formula. When E is a normed space, and F a Banach space, we have the following:

PROPOSITION 2.1 (Cauchy-Hadamard). The radius of uniform convergence of a power series $\sum_{m=0}^{\infty} P_m(x-\xi)$ from E into F about the point $\xi \in E$ is given by

$$r = \frac{1}{\limsup_{m\to\infty} \|P_m\|^{1/m}} .$$

REMARK 2.1: In this expression, r is taken to be 0 or ∞ if $\limsup_{m\to\infty} \|P_m\|^{1/m}$ is infinite or zero respectively.

PROOF: a) We consider first the case $\limsup_{m\to\infty} \|P_m\|^{1/m} = \infty$. Given $L \in \mathbb{R}$, $L > 0$ we denote by $M(L)$ the set of all $m \in \mathbb{N}$ for which $\|P_m\|^{1/m} > L$. Then $M(L)$ is an infinite set. For each $m \in M(L)$ choose $t_m \in E$ with $\|t_m\| = 1$, such that $\|P_m(t_m)\| > L^m$. Let $\rho = 1/L$, and let $x_m = \xi + \rho t_m$ for each $m \in M(L)$. Then

$$\|P_m(x_m-\xi)\| = \|P_m(\rho t_m)\| = \rho^m\|P_m(t_m)\| > \rho^m(1/\rho)^m = 1$$

for every $m \in M(L)$. This shows that the power series in question does not converge uniformly in $\bar{B}_\rho(\xi)$, since its general term does not converge uniformly to zero in this ball. Since L is an arbitrary positive real number, we conclude that the series does not converge uniformly in any ball $\bar{B}_\rho(\xi)$, $\rho \in \mathbb{R}$, $\rho > 0$. Hence the radius of convergence of the series is zero.

b) Suppose now that $\limsup_{m\to\infty} \|P_m\|^{1/m} = 0$.

Given $\rho \in \mathbb{R}$, $\rho > 0$, let $\varepsilon = \frac{1}{2\rho}$. For all but a finite number of values of m, $m \in \mathbb{N}$, we have $\|P_m\|^{1/m} \leq \varepsilon$. Let $t \in E$, $\|t\| \leq 1$, and let $x = \xi + \rho t$. Then

$$\|P_m(x-\xi)\| = \rho^m \|P_m(t)\| \leq \rho^m \varepsilon^m = 1/2^m$$

for all but finitely many m, $m \in \mathbb{N}$, and hence the given power series converges uniformly in $\bar{B}_\rho(\xi)$. Since ρ is an arbitrary positive real number, we conclude that the radius of convergence is infinite.

c) Finally, we consider the case

$$0 < \limsup_{m \to \infty} \|P_m\|^{1/m} = \Lambda < +\infty.$$

If $\rho \in \mathbb{R}$ and $0 < \rho < 1/\Lambda$, then $\Lambda < 1/\rho$, and so there exists $\theta \in \mathbb{R}$, $0 < \theta < 1$, such that $\Lambda < \theta/\rho$. Hence, for all but a finite number of values of m, $m \in \mathbb{N}$, we have $\|P_m\| \leq (\theta/\rho)^m$. Thus if $t \in E$, $\|t\| \leq 1$, and $x = \xi + \rho t$, then

$$\|P_m(x-\xi)\| = \rho^m \|P_m(t)\| \leq \rho^m (\theta/\rho)^m = \theta^m$$

for all but finitely many m, $m \in \mathbb{N}$, and hence the given power series converges uniformly in $\bar{B}_\rho(\xi)$. Therefore the radius of uniform convergence, r, satisfies $r \geq 1/\Lambda$.

If, on the other hand, we take $\rho > 1/\Lambda$, then $\Lambda > 1/\rho$, and so there exists an infinite subset M of $\mathbb{N}$ such that $\|P_m\| > (1/\rho)^m$ for $m \in M$. For every $m \in M$ there exists $t_m \in E$, $\|t_m\| = 1$, such that $\|P_m(t_m)\| > (1/\rho)^m$. Taking $x_m = \xi + \rho t_m$, we then have

$$\|P_m(x_m-\xi)\| = \rho^m\|P_m(t_m)\| > \rho^m(1/\rho)^m = 1$$

for every $m \in M$, and so the series does not converge uniformly in $\bar{B}_\rho(\xi)$. Thus $r \leq 1/\Lambda$.

Therefore $r = 1/\Lambda$. Q.E.D.

REMARK 2.2: For a power series $\sum_{m=0}^{\infty} P_m(x-\xi)$ from E into F about a point $\xi \in E$, the radius of convergence is unchanged if the norm on F is replaced by an equivalent norm; however, this radius can change if the norm on E is replaced by an equivalent one.

One can not, in general, replace the polynomials P_m by the corresponding A_m, $P_m = \hat{A}_m$, in the formula for the radius of convergence given in the preceeding proposition. This is illustrated by the following example:

EXAMPLE 2.1 Let E be the space ℓ^1 of absolutely summable sequences of complex numbers, with the norm $\|x\| = \sum_{m=1}^{\infty} |x_m|$ for $x = (x_1,x_2,\ldots) \in \ell^1$, and let $F = \mathbb{C}$. For $m = 1,2,\ldots$ we define $A_m\colon E^m \to F$ by

$$A_m(x^1,\ldots,x^m) = \frac{m^m}{m!} \sum_{\sigma\in S_m} x_1^{\sigma(1)}\ldots x_m^{\sigma(m)}$$

where $x^j = (x_1^j,\ldots,x_m^j,\ldots) \in E$, $j = 1,\ldots,m$.

It is easy to see that A_m is m-linear and symmetric, and, as in Example 1.2, we find that A_m is continuous, and that $\|A_m\| = \frac{m^m}{m!}$, $m \in \mathbb{N}$, $m \geq 1$. For $m = 0$, we take $A_o = 0$, and so $\|A_o\| = 0$.

The continuous m-homogeneous polynomial $P_m = \hat{A}_m$ associated with A_m is given by

$$P_m(x) = m^m x_1 \cdots x_m$$

for $x = (x_1, \ldots, x_m, \ldots) \in E$ and $m = 1,2,\ldots$. If $m = 0$, $P_m = 0$. Again following Example 1.2 we find that $\|P_m\| = 1$ for $m=1,2,\ldots$. Thus

$$\frac{1}{\limsup_{m\to\infty} \|P_m\|^{1/m}} = 1$$

but

$$\frac{1}{\limsup_{m\to\infty} \|A_m\|^{1/m}} = \frac{1}{\limsup_{m\to\infty} m/(m!)^{1/m}} = \frac{1}{e} .$$

COROLLARY 2.1 Let $\sum_{m=0}^{\infty} P_m(x-\xi) = \sum_{m=0}^{\infty} A_m(x-\xi)^m$ be a power series about $\xi \in E$. Then the following are equivalent:

a) The series is uniformly convergent.

b) The sequence $\{\|P_m\|^{1/m}\}$, $m \in \mathbb{N}$, is bounded.

c) The sequence $\{\|A_m\|^{1/m}\}$, $m \in \mathbb{N}$, is bounded.

PROOF: Statement a), that the radius of convergence is greater than zero, is equivalent to $\limsup_{m\to\infty} \|P_m\|^{1/m} \in \mathbb{R}$, and this in turn is equivalent to the assertion that the sequence $\{\|P_m\|^{1/m}\}$, $m \in \mathbb{N}$, is bounded. Hence a) and b) are equivalent.

The equivalence of b) and c) is a consequence of the relation

$$\|P_m\|^{1/m} \leq \|A_m\|^{1/m} \leq (m^m/m!)^{1/m} \|P_m\|^{1/m}$$

and the fact that the sequence $\{(m^m/m!)^{1/m}\}$, $m \in \mathbb{N}$, which converges to e, is bounded. Q.E.D.

PROPOSITION 2.2 Let $P_m \in \mathcal{P}_a({}^m E;F)$, $m \in \mathbb{N}$, $\xi \in E$, $\rho > 0$, and suppose that $\sum_{m=0}^{\infty} P_m(x-\xi) = 0$ for every $x \in B_\rho(\xi)$. Then $P_m = 0$ for every $m \in \mathbb{N}$.

PROOF: We begin by proving the following:

"If $\{u_m\}_{m\in\mathbb{N}}$ is a sequence of elements of F, $\delta > 0$, and $\sum_{m=0}^{\infty} \lambda^m u_m = 0$ for every $\lambda \in \mathbb{C}$, $|\lambda| \leq \delta$, then $u_m = 0$ for every $m \in \mathbb{N}$."

Taking $\lambda = 0$, we obtain $u_o = 0$. Suppose that $u_o = u_1 = \dots = u_{k-1} = 0$ for $k \geq 1$. We shall prove that $u_k = 0$, and our claim follows by induction: Since $\sum_{m=0}^{\infty} \delta^m u_m = 0$, $\lim_{m\to\infty} \delta^m u_m = 0$, and hence

1) $L = \sup_{m\in\mathbb{N}} \|u_m\| \delta^m < +\infty$.

By the induction hypothesis, $u_k = - \sum_{m=k+1}^{\infty} \lambda^{m-k} u_m$ for $\lambda \in \mathbb{C}$, $0 < |\lambda| \leq \delta$, and so by 1),

$$\|u_k\| \leq \sum_{m=k+1}^{\infty} |\lambda|^{m-k} \|u_m\| \leq \frac{L}{\delta^k} \frac{|\lambda|}{\delta - |\lambda|}$$

for $0 < |\lambda| < \delta$. Since

$$\lim_{\lambda\to 0} \frac{L}{\delta^k} \frac{|\lambda|}{\delta - |\lambda|} = 0,$$

it follows that $u_k = 0$.

Now let $t \in E$, $t \neq 0$, $\lambda \in \mathbb{C}$, and $x = \xi + \lambda t$. By hypothesis, $0 = \sum_{m=0}^{\infty} P_m(x-\xi) = \sum_{m=0}^{\infty} \lambda^m P_m(t)$ when $\|x-\xi\| =$

$= |\lambda| \|t\| < \rho$, that is, when $|\lambda| < \rho \|t\|^{-1}$. Applying the result proved above with $u_m = P_m(t)$ we find that $P_m(t) = 0$ for every $m \in \mathbb{N}$. Since this holds for all $t \in E$, we have $P_m = 0$ for every $m \in \mathbb{N}$. Q.E.D.

CHAPTER 3

HOLOMORPHIC MAPPINGS

DEFINITION 3.1 A mapping $f: U \to F$ is said to be holomorphic in U if for every $\xi \in U$ there exists a sequence $\{A_m\}_{m\in\mathbb{N}}$, $A_m \in \mathcal{L}_s(^mE;F)$, $m \in \mathbb{N}$, and a number $\rho \in \mathbb{R}$, $\rho > 0$, such that $B_\rho(\xi) \subset U$ and the power series $\sum_{m=0}^{\infty} A_m(x-\xi)^m$ from E into F converges uniformly to $f(x)$ in $B_\rho(\xi)$.

It follows as a consequence of Proposition 2.2 that if $f: U \to F$ is holomorphic in U, the sequence $\{A_m\}_{m\in\mathbb{N}}$, $A_m \in \mathcal{L}_s(^mE;F)$, associated to each $\xi \in U$ is unique.

If $P_m = \hat{A}_m$, then $\sum_{m=0}^{\infty} A_m(x-\xi)^m = \sum_{m=0}^{\infty} P_m(x-\xi)$ is called the Taylor series of f about the point $\xi \in U$, and we write

$$f(x) \cong \sum_{m=0}^{\infty} A_m(x-\xi)^m \qquad \text{or} \qquad f(x) \cong \sum_{m=0}^{\infty} P_m(x-\xi).$$

The set $\mathcal{H}(U;F)$ of mappings from U into F which are holomorphic in U forms a complex vector space, the operations of addition and multiplication by scalars being defined pointwise.

REMARK 3.1: We can also formulate a definition of holomorphy at a point: $f: U \to F$ is said to be holomorphic at $\xi \in U$

if there exists a sequence $\{A_m\}_{m\in\mathbb{N}}$, $A_m \in \mathcal{L}_s(^mE;F)$, $m \in \mathbb{N}$, and $\rho \in \mathbb{R}$, $\rho > 0$, such that $B_\rho(\xi) \subset U$ and the power series $\sum_{m=0}^{\infty} A_m(x-\xi)^m$ converges uniformly to $f(x)$ in $B_\rho(\xi)$.

It is shown in Nachbin [90] that when F is a Banach space and $f: U \to F$, the set of points in U at which f is holomorphic is open.

In the preceeding definitions of a holomorphic mapping, whether in an open subset or at a point, one can dispense with the condition that A_m, $m \in \mathbb{N}$, be continuous, if we assume that f is continuous on U. In fact, in this case there exists an open ball $B_\rho(\xi)$, for $\xi \in U$, where f is bounded and the power series $\sum_{m=0}^{\infty} A_m(x-\xi)^m = \sum_{m=0}^{\infty} P_m(x-\xi)$ converges uniformly to f. This implies that each P_m, $m \in N$, is bounded in $B_\rho(\xi)$. Therefore, by Proposition 1.7 we have that P_m (and hence also A_m) is continuous for every $m \in \mathbb{N}$.

We note too that the set $\mathcal{H}(U;F)$ remains unchanged if the norms on E and F are replaced by equivalent norms.

DEFINITION 3.2 Let $f \in \mathcal{H}(U;F)$, $\xi \in U$, and let $\sum_{m=0}^{\infty} A_m(x-\xi)^m = \sum_{m=0}^{\infty} P_m(x-\xi)$ be the Taylor series of f at ξ, where $P_m = \hat{A}_m$, $m \in \mathbb{N}$. Then

a) $d^m f(\xi) = m!A_m \in \mathcal{L}_s(^mE;F)$ and

b) $\hat{d}^m f(\xi) = m!\hat{A}_m = m!P_m \in \mathcal{P}(^mE;F)$

are called the differential of order m of f at ξ. In a) the differential is viewed as a continuous symmetric m-linear mapping, and in b) as a continuous m-homogeneous polynomial. Note that $\hat{d}^m f(\xi)$ is an abbreviated notation for $\widehat{d^m f(\xi)}$.

The Taylor series of f at ξ can now be wrriten as

$$f(x) \simeq \sum_{m=0}^{\infty} \frac{1}{m!} d^m f(\xi)(x-\xi)^m$$

or

$$f(x) \simeq \sum_{m=0}^{\infty} \frac{1}{m!} \hat{d}^m f(\xi)(x-\xi).$$

$\tau_{m,f,\xi}(x) = \sum_{k=0}^{m} \frac{1}{k!} \hat{d}^k f(\xi)(x-\xi)$ is the Taylor polynomial of degree m of f at ξ.

If $f \in \mathcal{H}(U;F)$, we can consider the differentials of f or order m, $m \in \mathbb{N}$, as mappings defined on U:

$$d^m f\colon \xi \in U \mapsto d^m f(\xi) \in \mathcal{L}_s({}^m E;F)$$

$$\hat{d}^m f\colon \xi \in U \mapsto \hat{d}^m f(\xi) \in \mathcal{P}({}^m E;F).$$

We can go a step further and define the differential operators of order m on $\mathcal{H}(U;F)$:

$$d^m\colon f \in \mathcal{H}(U;F) \mapsto d^m f \in \mathcal{F}(U;\mathcal{L}_s({}^m E;F))$$

$$\hat{d}^m\colon f \in \mathcal{H}(U;F) \mapsto \hat{d}^m f \in \mathcal{F}(U;\mathcal{P}({}^m E;F)).$$

DEFINITION 3.3 Let $m > 0$, let $E_1,\ldots,E_m$ and F be normed spaces, and $A \in \mathcal{L}_a(E_1,\ldots,E_m;F)$. For $h \in \mathbb{N}$, $0 \leq h \leq m$, and $x_1 \in E_1,\ldots,x_h \in E_h$, $A(x_1,\ldots,x_h)$ is defined as follows:

1) If $h = m$, $A(x_1,\ldots,x_h) = A(x_1,\ldots,x_m) \in F$ is the value of the mapping A at the m-tuple $(x_1,\ldots,x_m)$.

2) If $h < m$, $A(x_1,\ldots,x_h)$ is the mapping from

$E_{h+1} \times \ldots \times E_m$ into F given by

$$A(x_1,\ldots,x_h)(x_{h+1},\ldots,x_m) = A(x_1,\ldots,x_h,x_{h+1},\ldots,x_m) \in F$$

for $(x_{h+1},\ldots,x_m) \in E_{h+1} \times \ldots \times E_m$.

It is easy to see that

a) $A(x_1,\ldots,x_h)$ is an $(m-h)$-linear mapping of $E_{h+1} \times \ldots \times E_m$ into F, which is continuous if A is continuous.

b) If $E_1 = \ldots = E_m = E$ and $A \in \mathcal{L}_{as}(^mE;F)$ or, more generally, if $E_{h+1} = \ldots = E_m = E$ and A is symmetric in the last $m-h$ variables, then $A(x_1,\ldots,x_h) \in \mathcal{L}_{as}(^{m-h}E;F)$.

If $E_1 = \ldots = E_h = E$ and $x_1 = \ldots = x_h = x \in E$ we write $A(x_1,\ldots,x_h) = Ax^h$ for every $h \in \mathbb{N}$, $1 \leq h \leq m$, and for $h = 0$ we set $Ax^o = A$. With this convention, if $m \in \mathbb{N}$, $A \in \mathcal{L}_a(^mE;F)$ and $x \in E$, then $Ax^h \in \mathcal{L}_a(^{m-h}E;F)$ is defined for every $h \in \mathbb{N}$, $0 \leq h \leq m$.

If $A \in \mathcal{L}_a(^mE;F)$, $1 \leq r \leq m$, $x_1,\ldots,x_r \in E$ and $h_1,\ldots,h_r \in \mathbb{N}$, $1 \leq h_1 + \ldots + h_r = h \leq m$, we define $Ax_1^{h_1} \ldots x_r^{h_r}$ to be the mapping

$$A(\overbrace{x_1,\ldots,x_1}^{h_1},\overbrace{x_2,\ldots,x_2}^{h_2},\ldots,\overbrace{x_r,\ldots,x_r}^{h_r}) \in \mathcal{L}_a(^{m-h}E;F).$$

DEFINITION 3.4 A mapping $f: E \to F$ is said to be entire if f is holomorphic in E. $\mathcal{H}(E;F)$ denotes the complex vector space of entire mappings of E into F, with the usual operations.

PROPOSITION 3.1 $\mathcal{P}(E;F)$ is a subset of $\mathcal{H}(E;F)$. More precisely, if $m \in \mathbb{N}$, $A \in \mathcal{L}_s(^mE;F)$, $P = \hat{A}$ and $\xi \in E$, then

$$\frac{1}{k!} d^kP(\xi) = \binom{m}{k} A\xi^{m-k}$$

if $k \in \mathbb{N}$, $0 \leq k \leq m$, and

$$\frac{1}{k!} d^kP(\xi) = 0$$

if $k \in \mathbb{N}$, $k > m$.

PROOF: It suffices to show that $\mathcal{P}(^mE;F) \subset \mathcal{H}(E;F)$ for every $m \in \mathbb{N}$. This is obvious in the cases $m = 0$ and $m = 1$. For $m \geq 2$, let $A \in \mathcal{L}_s(^mE;F)$ and $P = \hat{A}$. Then for every $\xi \in E$ we have

$$P(x) = \sum_{k=0}^{m} \binom{m}{k} A\xi^{m-k}(x-\xi)^k$$

for every $x \in E$. This shows that P is holomorphic at every $\xi \in E$, and that the radius of convergence of the Taylor series of P at every $\xi \in E$ is infinite, since this series is finite and hence converges uniformly to $P(x)$ in $\bar{B}_\rho(\xi)$ for every $\rho \in \mathbb{R}$, $\rho > 0$. The expressions for the differentials of P at any $\xi \in E$ follow directly from the relation above.

Q.E.D.

REMARK 3.2: In the case in which E has finite dimension and $f \in \mathcal{H}(E;F)$, the radius of convergence of the Taylor series of f at ξ is infinite for every $\xi \in E$. If the dimension of E is infinite, the radius of convergence of the Taylor series of $f \in \mathcal{H}(E;F)$ is either finite at every point of E or infinite at every point of E. The second possibility characterises the entire functions of bounded type, which we shall consider later.

COROLLARY 3.1 Let $m \in \mathbb{N}$, $P \in \mathcal{P}(^mE;F)$ and $k \in \mathbb{N}$, $0 \leq k \leq m$. Then

$$d^kP \in \mathcal{P}(^{m-k}E; \mathcal{L}_s(^kE;F))$$

and

$$\hat{d}^kP \in \mathcal{P}(^{m-k}E; \mathcal{P}(^kE;F)).$$

PROPOSITION 3.2 Let E, F and G be normed spaces, U a non-empty open subset of E, $f \in \mathcal{H}(U;F)$ and $t \in \mathcal{L}(F;G)$. Then $t \circ f \in \mathcal{H}(U;G)$ and $d^m(t \circ f)(\xi) = t \circ (d^m f(\xi))$ for every $\xi \in U$ and $m \in \mathbb{N}$.

PROOF: If $\xi \in U$, there exists $\rho \in \mathbb{R}$, $\rho > 0$ and a continuous symmetric m-linear mapping $d^m f(\xi)$ for every $m \in \mathbb{N}$ such that $B_\rho(\xi) \subset U$ and

$$1) \qquad f(x) = \sum_{m=0}^{\infty} \frac{1}{m!} d^m f(\xi)(x-\xi)^m$$

uniformly in $\bar{B}_\rho(\xi)$. Since $t: F \to G$ is linear and continuous, and hence is uniformly continuous in F, 1) implies that

$$2) \qquad t\{f(x)\} = t\{\sum_{m=0}^{\infty} \frac{1}{m!} d^m f(\xi)(x-\xi)^m\}$$

$$= \sum_{m=0}^{\infty} \frac{1}{m!} \{t \circ (d^m f(\xi))\}(x-\xi)^m$$

uniformly in $\bar{B}_\rho(\xi)$. It is easily seen, using Proposition 1.6, that $t \circ (d^m f(\xi)) \in \mathcal{L}_s(^mE;G)$ for every $m \in \mathbb{N}$. Therefore, by 2), $t \circ f \in \mathcal{H}(U;G)$, and by the uniqueness of the Taylor series we have $d^m(t \circ f)(\xi) = t \circ (d^m f(\xi))$ for every $\xi \in U$ and $m \in \mathbb{N}$.

Q.E.D.

CHAPTER 4

THE CAUCHY INTEGRAL FORMULAS

REMARK 4.1: In this chapter we make use of some properties of integrals of functions defined on a subset of $\mathbb{R}$ or $\mathbb{C}$ with values in a real or complex Banach space (see Dieudonné [28] and Hille [57]).

PROPOSITION 4.1 (The Cauchy Integral Formula).

Let $f \in \mathcal{H}(U;F)$, $\xi \in U$, $x \in U$ and $\rho \in \mathbb{R}$, $\rho > 1$ such that $(1-\lambda)\xi + \lambda x \in U$ for every $\lambda \in \mathbb{C}$, $|\lambda| \leq \rho$. Then

$$(*) \qquad f(x) = \frac{1}{2\pi i} \int_{|\lambda|=\rho} \frac{f[(1-\lambda)\xi + \lambda x]}{\lambda - 1} \, d\lambda .$$

REMARK 4.2: Although F is not necessarily complete, the existence of the integral in (*) is guaranteed, since we may consider f as taking its values in the completion $\hat{F}$ of F, identifying F with its image in $\hat{F}$ under the natural isometric inclusion I. The relation (*) will be proved for $I \circ f$, and as $(I \circ f)(x) = f(x) \in F$, it follows that the integral in (*) is an element of F.

PROOF: By the preceeding remark, it is sufficient to consider the case in which F is a Banach space. We shall make use the following well-known result:

a) Let V be a non-empty open subset of $\mathbb{C}$, $z \in V$, $\delta \in \mathbb{R}$, $\delta > 0$, such that $\bar{D}_\delta(z) \subset V$, $\bar{D}_\delta(z)$ being the usual closed ball of radius ρ and center z in $\mathbb{C}$, and let $g \in \mathcal{H}(V;F)$ and $\tau \in D_\delta(z)$. Then

$$g(\tau) = \frac{1}{2\pi i} \int_{|\lambda - z| = \delta} \frac{g(\lambda)}{\lambda - \tau} \, d\lambda .$$

With ξ, x, ρ as in the statement of the proposition, $V = \{\lambda \in \mathbb{C} : (1-\lambda)\xi + \lambda x \in U\}$ is an open subset of $\mathbb{C}$, $\bar{D}_\rho(0) \subset V$ and $1 \in D_\rho(0)$. If $g: V \to F$ is defined by $g(\lambda) = f[(1-\lambda)\xi + \lambda x]$ for $\lambda \in V$, then it is easy to see that $g \in \mathcal{H}(V;F)$. Applying a) to g, with $\tau = 1$, we obtain

$$f(x) = g(1) = \frac{1}{2\pi i} \int_{|\lambda| = \rho} \frac{g(\lambda)}{\lambda - 1} \, d\lambda$$

$$= \frac{1}{2\pi i} \int_{|\lambda| = \rho} \frac{f[(1-\lambda)\xi + \lambda x]}{\lambda - 1} \, d\lambda . \qquad \text{Q.E.D.}$$

PROPOSITION 4.2 (The Cauchy Integral Formulas).

Let $f \in \mathcal{H}(U;F)$, $\xi \in U$, $x \in E$ and $\rho \in \mathbb{R}$, $\rho > 0$ such that $\xi + \lambda x \in U$ for every $\lambda \in \mathbb{C}$, $|\lambda| \le \rho$. Then

$$\frac{1}{m!} \hat{d}^m f(\xi)(x) = \frac{1}{m!} d^m f(\xi) x^m = \frac{1}{2\pi i} \int_{|\lambda| = \rho} \frac{f(\xi + \lambda x)}{\lambda^{m+1}} \, d\lambda$$

for every $m \in \mathbb{N}$.

PROOF: As in the preceeding proposition, it is sufficient to consider the case in which F is a Banach space.

We shall make use of the following well-known result:

a) Let V be a non-empty open subset of $\mathbb{C}$, $g \in \mathcal{H}(V;F)$, $r, R \in \mathbb{R}$, $0 < r < R$, $a \in V$, such that $\{\lambda \in \mathbb{C} : r \le |\lambda - a| \le R\} \subset V$. Then

$$\int_{|\lambda-a|=r} g(\lambda)d\lambda = \int_{|\lambda-a|=R} g(\lambda)d\lambda .$$

Now $V = \{\lambda \in \mathbb{C}: \lambda \neq 0, \xi+\lambda x \in U\}$ is a non-empty open subset of $\mathbb{C}$, and $V \supset \bar{B}_\rho(0) - \{0\}$. If $g: V \to V$ is defined by $g(\lambda) = \frac{f(\xi+\lambda x)}{\lambda^{m+1}}$, $\lambda \in V$, it is easy to see that $g \in \mathcal{H}(V;F)$. If $0 < \epsilon < \rho$ then $\{\lambda \in \mathbb{C} : \epsilon \leq |\lambda| \leq \rho\} \subset V$ and so, by a), we have

$$(*) \qquad \int_{|\lambda|=\rho} \frac{f(\xi+\lambda x)}{\lambda^{m+1}} d\lambda = \int_{|\lambda|=\epsilon} \frac{f(\xi+\lambda x)}{\lambda^{m+1}} d\lambda$$

for every $m \in N$.

The Taylor series of f at ξ, $\sum_{\ell=0}^{\infty} P_\ell(Z-\xi)$, converges uniformly to $f(z)$ in a ball $B_\sigma(\xi)$, $\sigma \in \mathbb{R}$, $\sigma > 0$. With $x \neq 0$, choose $\epsilon \in \mathbb{R}$, $0 < \epsilon < \rho$, sufficiently small so that the series $\sum_{\ell=0}^{\infty} P_\ell(Z-\xi)$ converges uniformly to $f(z)$ in the closed ball with centre ξ and radius $\epsilon\|x\|$. Then if $z = \xi + \lambda x$ we have $\|z-\xi\| = |\lambda|\|x\| \leq \epsilon\|x\|$ for $\lambda \in \mathbb{C}$, $|\lambda| \leq \epsilon$. Therefore, from (*), we have

$$\int_{|\lambda|=\rho} \frac{f(\xi+\lambda x)}{\lambda^{m+1}} d\lambda = \int_{|\lambda|=\epsilon} \{\sum_{\ell=0}^{\infty} P_\ell(z-\xi) \frac{1}{\lambda^{m+1}}\} d\lambda$$

$$= \int_{|\lambda|=\epsilon} \{\sum_{\ell=0}^{\infty} \lambda^\ell P_\ell(x) \frac{1}{\lambda^{m+1}}\} d\lambda .$$

Since the series converges uniformly in $\{\lambda\in\mathbb{C}; |\lambda|=\epsilon\}$, this yields

$$\int_{|\lambda|=\rho} \frac{f(\xi+\lambda x)}{\lambda^{m+1}}\, d\lambda = \sum_{\ell=0}^{\infty} P_\ell(x) \int_{|\lambda|=\varepsilon} \frac{1}{\lambda^{m+1-\ell}}\, d\lambda$$

$$= 2\pi i\, P_m(x) = 2\pi i \frac{1}{m!} \hat{d}^m f(\xi)(x).$$

For $x = 0$, the proposition is trivial. Q.E.D.

REMARK 4.3: In the case $m = 0$, we obtain the following corollary: if $f \in \mathcal{H}(U,F)$, $\xi \in U$, $x \in E$, and $\rho \in \mathbb{R}$, $\rho > 0$ such that $\xi+\lambda x \in U$ for $\lambda \in \mathbb{C}$, $|\lambda| \le \rho$, then

$$f(\xi) = \frac{1}{2\pi i} \int_{|\lambda|=\rho} \frac{f(\xi+\lambda x)}{\lambda}\, d\lambda .$$

The following proposition is a generalization of Propostion 4.2:

PROPOSITION 4.3 Let $f \in \mathcal{H}(U;F)$, $\xi \in U$, $m \in \mathbb{N}$, $m \ge 1$, $k \in \mathbb{N}$, $1 \le k \le m$, $(x_1,\dots,x_k)$ a k-tuple of elements of E, and $(\rho_1,\dots,\rho_k)$ a k-tuple of strictly positive real numbers such that $\xi + \sum_{j=1}^{k} \lambda_j x_j \in U$ for every $(\lambda_1,\dots,\lambda_k) \in \mathbb{C}^k$ with $|\lambda_j| \le \rho_j$, $j = 1,\dots,k$. Then if $n = (n_1,\dots,n_k) \in \mathbb{N}^k$ is such that $|n| = \sum_{j=1}^{k} n_j = m$,

$$\frac{1}{n_1!\dots n_k!} d^m f(\xi) x_1^{n_1} \dots x_k^{n_k} =$$

$$= \frac{1}{(2\pi i)^k} \int_{\substack{|\lambda_j|=\rho_j \\ 1\le j\le k}} \frac{f(\xi + \sum_{j=1}^{k} \lambda_j x_j)}{\lambda_1^{n_1+1} \dots \lambda_k^{n_k+1}}\, d\lambda_1 \dots d\lambda_k .$$

REMARK 4.4: In the case $k = 1$ the expression which appear in Proposition 4.3 reduces to that of Proposition 4.2, the expression for the coefficient of order m of the Taylor series taking the form of a homogeneous polynomial, and the integral being a simple contour integral. In the other extreme case, $k = m$, we obtain

$$d^m f(\xi)(x_1,\ldots,x_m) = \frac{1}{(2\pi i)^m} \int_{\substack{|\lambda_j|=\rho_j \\ 1\leq j\leq m}} \frac{f(\xi + \sum_{j=1}^{m} \lambda_j x_j)}{\lambda_1^2 \cdots \lambda_m^2} d\lambda_1 \cdots d\lambda_m .$$

The following proposition is a consequence of the Cauchy integral formulas:

PROPOSITION 4.4 (The Cauchy Inequalities). Let $f \in \mathcal{H}(U,F)$, $\xi \in U$ and $\rho \in \mathbb{R}$, $\rho > 0$ such that $\bar{B}_\rho(\xi) \subset U$. Then for every $m \in \mathbb{N}$,

$$\left\|\frac{1}{m!} \hat{d}^m f(\xi)\right\| \leq \frac{1}{\rho^m} \sup_{\|x-\xi\|=\rho} \|f(x)\| .$$

PROOF: Let $t \in E$, $\|t\| = 1$, and let $x = \xi + \rho t \in \bar{B}_\rho(\xi) \subset U$. Then, by Proposition 4.2,

$$\frac{1}{m!} \hat{d}^m f(\xi)(t) = \frac{1}{2\pi i} \int_{|\lambda|=\rho} \frac{f(\xi+\lambda t)}{\lambda^{m+1}} d\lambda ,$$

and hence

$$\left\|\frac{1}{m!} \hat{d}^m f(\xi)(t)\right\| \leq \frac{1}{2\pi} \sup_{|\lambda|=\rho} \|f(\xi+\lambda t)\| \frac{1}{\rho^{m+1}} 2\pi\rho$$

$$\leq \frac{1}{\rho^m} \sup_{\|x-\xi\|=\rho} \|f(x)\| .$$

Since this holds for every $t \in E$ with $\|t\| = 1$ we have

$$\left\|\frac{1}{m!}\hat{d}^m f(\xi)\right\| \leq \frac{1}{\rho^m} \sup_{\|x-\xi\|=\rho} \|f(x)\|$$

Q.E.D.

REMARK 4.5: Taking into account the inequalities

$$\|\hat{A}\| \leq \|A\| \leq \frac{m^m}{m!}\|\hat{A}\|,$$

valid for every $A \in \mathcal{L}_s({}^m E;F)$, we can write the Cauchy inequalities in the form

$$(*) \qquad \left\|\frac{1}{m!} d^m f(\xi)\right\| \leq \frac{m^m}{m!} \cdot \frac{1}{\rho^m} \cdot \sup_{\|x-\xi\|=\rho} \|f(x)\|,$$

where $f \in \mathcal{H}(U;F)$ and $m \in \mathbb{N}$.

The constants $\frac{1}{\rho^m}$ and $\frac{1}{\rho^m} \cdot \frac{m^m}{m!}$ which appear respectively in the Cauchy inequalities and in (*) are the least possible universal constants in these inequalities.

PROPOSITION 4.5 Let $f \in \mathcal{H}(U;F)$, $\xi \in U$, $x \in U$ and $\rho \in \mathbb{R}$, $\rho > 1$, such that $(1-\lambda)\xi + \lambda x \in U$ for every $\lambda \in \mathbb{C}$, $|\lambda| \leq \rho$. Then for every $m \in \mathbb{N}$,

$$f(x) - \tau_{m,f,\xi}(x) = \frac{1}{2\pi i} \int_{|\lambda|=\rho} \frac{f[(1-\lambda)\xi + \lambda x]}{\lambda^{m+1}(\lambda-1)} d\lambda,$$

where

$$\tau_{m,f,\xi}(x) = \sum_{k=0}^{m} \frac{1}{k!} \hat{d}^k f(\xi)(x-\xi).$$

PROOF: For $\lambda \in \mathbb{C}$, $\lambda \neq 0$ and $\lambda \neq 1$, we have

$$(*) \qquad \frac{1}{\lambda-1} = \sum_{k=0}^{m} \frac{1}{\lambda^{k+1}} + \frac{1}{\lambda^{m+1}(\lambda-1)}$$

for every $m \in \mathbb{N}$.

Multiplying (*) by $\frac{1}{2\pi i} f[(1-\lambda)\xi+\lambda x]$ and integrating the resulting equation over the circle $|\lambda| = \rho$, we obtain

$$(**) \qquad \frac{1}{2\pi i}\int_{|\lambda|=\rho} \frac{f[(1-\lambda)\xi+\lambda x]}{\lambda-1}\, d\lambda =$$

$$\sum_{k=0}^{m} \frac{1}{2\pi i}\int_{|\lambda|=\rho} \frac{f[(1-\lambda)\xi+\lambda x]}{\lambda^{k+1}}\, d\lambda + \frac{1}{2\pi i}\int_{|\lambda|=\rho} \frac{f[(1-\lambda)\xi+\lambda x]}{\lambda^{m+1}(\lambda-1)}\, d\lambda .$$

Applying Proposition 4.1 to the left hand side of (**) and Proposition 4.2, with $(x-\xi)$ in place of x, to the first term of the right hand side (noting that $(1-\lambda)\xi + \lambda x = \xi + \lambda(x-\xi)$), we obtain the desired relation. Q.E.D.

COROLLARY 4.1 Under the hypotheses of Proposition 4.5 we have the following estimate for the norm of the Taylor remainder of order m for f at ξ:

$$\|f(x) - \tau_{m,f,\xi}(x)\| \leq \frac{1}{\rho^m(\rho-1)} \sup_{|\lambda|=\rho} \|f[(1-\lambda)\xi+\lambda x]\| .$$

PROOF: This follows immediately from Proposition 4.5, using the usual inequality for the norm of an integral and the fact that $\inf_{|\lambda|=\rho} |\lambda-1| = \rho-1$. Q.E.D.

COROLLARY 4.2 Let $f \in \mathcal{H}(U;F)$, $m \in \mathbb{N}$, $\xi \in U$ and $r \in \mathbb{R}$, $r > 0$, such that $\bar{B}_r(\xi) \subset U$, and let $x \in B_r(\xi)$. Then

$$\|f(x) - \sum_{\ell=0}^{m} \frac{1}{\ell!} d^\ell f(\xi)(x-\xi)^\ell\| \leq$$

$$\frac{\|x-\xi\|^{m+1}}{r^m(r-\|x-\xi\|)} \cdot \sup_{\|t-\xi\|=r} \|f(t)\| .$$

PROOF: If $x = \xi$, the inequality is trivial. For $x \neq \xi$, let $\rho = \frac{r}{\|x-\xi\|}$. Then $\rho > 1$ and $(1-\lambda)\xi + \lambda x \in U$ if $\lambda \in \mathbb{C}$, $|\lambda| \leq \rho$. Thus, by Proposition 4.5,

$$f(x) - \sum_{\ell=0}^{m} \frac{1}{\ell!} d^{\ell} f(\xi)(x-\xi)^{\ell} = \frac{1}{2\pi i} \int_{|\lambda|=\rho} \frac{f[(1-\lambda)\xi+\lambda x]}{\lambda^{m+1}(\lambda-1)} d\lambda .$$

Applying the usual inequality for the norm of an integral, and the fact that

$$\|\xi+\lambda(x-\xi)-\xi\| = |\lambda|\cdot\|x-\xi\| = \rho\|x-\xi\| = r \quad \text{if} \quad |\lambda| = \rho,$$

which implies

$$\sup_{|\lambda|=\rho} \|f[(1-\lambda)\xi+\lambda x]\| \leq \sup_{\|t-\xi\|=r} \|f(t)\|,$$

we obtain:

$$\|f(x) - \sum_{\ell=0}^{m} \frac{1}{\ell!} d^{\ell} f(\xi)(x-\xi)^{\ell}\| \leq \frac{1}{\rho^{m}(\rho-1)} \sup_{\|t-\xi\|=r} \|f(t)\| .$$

The corollary now follows when $\frac{r}{\|x-\xi\|}$ is substituted for ρ.

Q.E.D.

REMARK 4.6: Consider the vector space $C(U;F)$ of continuous mappings of U into F. For each compact subset K of U, the mapping

$$p_K: C(U;F) \to \mathbb{R},$$

defined by

$$p_K(f) = \sup_{x\in K} \|f(x)\|; \quad f \in C(U;F),$$

is a seminorm on $C(U;F)$.

The separated locally convex topology defined on $C(U;F)$

by the family of seminorms $\{p_K\colon K$ compact, $K \subset U\}$ is known as the compact-open topology.

PROPOSITION 4.6 $\mathcal{H}(U;F)$ is a vector subspace of $C(U;F)$. If F is complete, and $C(U;F)$ carries the compact-open topology, then $\mathcal{H}(U;F)$ is a closed vector subspace of $C(U;F)$, and hence is complete in the induced topology.

PROOF: If $f \in \mathcal{H}(U;F)$, let $\xi \in U$, and $\rho \in R$, $\rho > 0$ such that the Taylor series of f at ξ converges to f uniformly in $B_\rho(\xi)$. By the Corollary 2.1, there exists $C > 0$ such that $\|P_m\|^{1/m} = \|\frac{1}{m!}\hat{d}^m f(\xi)\|^{1/m} \le C$ for every $m \in \mathbb{N}$, $m \ge 1$. Since $P_o(x-\xi) = P_o = f(\xi)$, and $\sum_{m=0}^{\infty} P_m(x-\xi)$ converges uniformly to $f(x)$ in $B_\rho(\xi)$, we have

$$\|f(x)-f(\xi)\| \le \sum_{m=1}^{\infty} C^m\|x-\xi\|^m = \frac{C\|x-\xi\|}{1-C\|x-\xi\|}$$

if $\|x-\xi\| < \rho$ and $\rho\cdot C < 1$. It follows that f is continuous at ξ, and therefore $\mathcal{H}(U;F) \subset C(U;F)$.

Suppose now that F is a Banach space. $C(U;F)$ is complete by the following result: "If X is a metrizable or locally compact space, and Y a Banach space, then $C(X;Y)$ is complete in the compact-open topology" (see Bourbaki [19], Topologie General, Chapter 10). To prove that $\mathcal{H}(U;F)$ is closed in $C(U;F)$ for the compact-open topology, we take $f \in \overline{\mathcal{H}(U;F)}$ and show that $f \in \mathcal{H}(U;F)$.

Let $\xi \in U$. We set $A_o = f(\xi)$, and for each $m \in \mathbb{N}$, $m \ge 1$, we define $A_m\colon E^m \to F$ in the following manner:

if $(x_1,\ldots,x_m) \in E^m$, there exists an m-tuple $(\rho_1,\ldots,\rho_m)$ of strictly positive real numbers, which we shall call an m-tuple associated to $(x_1,\ldots,x_m)$, such that

$$\xi + \sum_{j=1}^{m} \lambda_j x_j \in U \quad \text{if} \quad \lambda_j \in \mathbb{C}, \quad |\lambda_j| \leq \rho_j, \quad j = 1,\ldots,m.$$

We then define

$$\text{a)} \quad A_m(x_1,\ldots,x_m) = \frac{1}{m!(2\pi i)^m} \int_{\substack{|\lambda_j|=\rho_j \\ 1\leq j\leq m}} \frac{f(\xi + \sum_{j=1}^{m} \lambda_j x_j)}{(\lambda_1\cdots\lambda_m)^2} d\lambda_1\cdots d\lambda_m$$

We note first that the integral in a) is defined since f is continuous and F is complete. We show next that the value of this integral is independent of the choice of the m-tuple $(\rho_1,\ldots,\rho_m)$ associated to $(x_1,\ldots,x_m)$. This is clear for $f \in \mathcal{H}(U;F)$ since, in this case,

$$A_m(x_1,\ldots,x_m) = \frac{1}{m!} d^m f(\xi)(x_1,\ldots,x_m)$$

as was noted in Remark 4.4. Consider the following mapping: fixing $(x_1,\ldots,x_m) \in E^m$, and an associated m-tuple $(\rho_1,\ldots,\rho_m)$, let $T_{\rho_1,\ldots,\rho_m;x_1,\ldots,x_m} : C(U;F) \to F$ be given by

$$T_{\rho_1,\ldots,\rho_m;x_1,\ldots,x_m}(g) =$$

$$= \frac{1}{m!(2\pi i)^m} \int_{\substack{|\lambda_j|=P_j \\ 1\leq j\leq m}} \frac{g(\xi + \sum_{i=1}^{m} \lambda_j x_j)}{(\lambda_1\cdots\lambda_m)^2}, \qquad g \in C(U;F).$$

This mapping is easily seen to be continuous, and by Remark 4.4, if $(\rho_1,\ldots,\rho_m)$ and $(\rho'_1,\ldots,\rho'_m)$ are two m-tuples associated to $(x_1,\ldots,x_m)$, then

$$T_{\rho_1,\ldots,\rho_m;x_1,\ldots,x_m}\Big|_{\mathcal{H}(U;F)} = T_{\rho'_1,\ldots,\rho'_m;x_1,\ldots,x_m}\Big|_{\mathcal{H}(U;F)}.$$

Since $T_{\rho_1,\ldots,\rho_m;x_1,\ldots,x_m}$ and $T_{\rho'_1,\ldots,\rho'_m;x_1,\ldots,x_m}$ are continuous, it follows that

$$T_{\rho_1,\ldots,\rho_m;x_1,\ldots,x_m}\Big|_{\overline{\mathcal{H}(U;F)}} = T_{\rho'_1,\ldots,\rho'_m;x_1,\ldots,x_m}\Big|_{\overline{\mathcal{H}(U;F)}}.$$

Therefore A_m is well-defined for every $m \in \mathbb{N}$.

A_m is symmetric and m-linear. This is proved using the continuity of the mapping $T_{\rho_1,\ldots,\rho_m;x_1,\ldots,x_m}$; we shall prove only that A_m is additive in each variable. Let $x_1,x_2,\ldots,x_j,x'_j,\ldots,x_m \in E$ and $g \in \mathcal{H}(U;F)$. Choosing an m-tuple $(\rho_1,\ldots,\rho_m)$ which is associated to each of the three elements $(x_1,\ldots,x_j,\ldots,x_m)$, $(x_1,\ldots,x'_j,\ldots,x_m)$ and $(x_1,\ldots,x_j + x'_j,\ldots,x_m)$ of E^m, we have

$$T_{\rho_1,\ldots,\rho_m;x_1,\ldots,x_j+x'_j,\ldots,x_m}(g) =$$

$$T_{\rho_1,\ldots,\rho_m;x_1,\ldots,x_j,\ldots,x_m}(g) + T_{\rho_1,\ldots,\rho_m;x_1,\ldots,x'_j,\ldots,x_m}(g).$$

Therefore

b) $$T_{\rho_1,\ldots,\rho_m;x_1,\ldots,x_j+x'_j,\ldots,x_m}\Big|_{\mathcal{H}(U;F)} = T_{\rho_1,\ldots,\rho_m;x_1,\ldots,x_j,\ldots,x_m}\Big|_{\mathcal{H}(U;F)} + T_{\rho_1,\ldots,\rho_m;x_1,\ldots,x'_j,\ldots,x_m}\Big|_{\mathcal{H}(U;F)}.$$

From b), and the continuity of the three mappings which appear, we have

$$T_{\rho_1,\dots,\rho_m;x_1,\dots,x_j+x'_j,\dots,x_m}\Big|_{\overline{\mathcal{H}(U;F)}} =$$

$$\{T_{\rho_1,\dots,\rho_m;x_1,\dots,x_j,\dots,x_m}+T_{\rho_1,\dots,\rho_m;x_1,\dots,x'_j,\dots,x_m}\}\Big|_{\overline{\mathcal{H}(U;F)}}$$

Therefore, as $f \in \overline{\mathcal{H}(U;F)}$, we have

$$A_m(x_1,\dots,x_j+x'_j,\dots,x_m) = T_{\rho_1,\dots,\rho_m;x_1,\dots,x_j+x'_j,\dots,x_m}(f) =$$

$$T_{\rho_1,\dots,\rho_m;x_1,\dots,x_j,\dots,x_m}(f)+T_{\rho_1,\dots,\rho_m;x_1,\dots,x'_j,\dots,x_m}(f) =$$

$$A_m(x_1,\dots,x_j,\dots,x_m) + A_m(x_1,\dots,x'_j,\dots,x_m).$$

Analogous considerations show that A_m is homogeneous in each variable, and symmetric. Thus $A_m \in \mathcal{L}_{as}(^mE;F)$ for every $m \in \mathbb{N}$.

We prove next that A_m, $m \geq 1$, is continuous. Fix a real number $\rho > 0$ such that 1) $0 < \rho < 1$, 2) $\bar{B}_\rho(\xi) \subset U$, and 3) there exists a real number $M > 0$ such that $\sup_{\|t-\xi\|\leq\rho} \|f(t)\| \leq M$, and let $(\rho_1,\dots,\rho_m)$ be an m-tuple of strictly positive real numbers with $\rho_1 + \dots + \rho_m = \rho$. Since $\bar{B}_\rho(\xi) \subset U$, if $(x_1,\dots,x_m) \in E^m$ with $\|x_1\| = \dots = \|x_m\| = 1$, then $\xi + \sum_{j=1}^{m} \lambda_j x_j \in U$ for every $(\lambda_1,\dots,\lambda_m) \in C^m$ with $|\lambda_j| \leq \rho_j$, $1 \leq j \leq m$. Therefore the m-tuple $(\rho_1,\dots,\rho_m)$ is associated to every $(x_1,\dots,x_m) \in E^m$ for which $\|x_j\| = 1$, $j = 1,\dots,m$. Thus for $(x_1,\dots,x_m) \in E^m$, $\|x_1\| = \dots = \|x_m\| = 1$, we have

$$\|A_m(x_1,\ldots,x_m)\| \leq$$

$$\frac{1}{m!}\frac{1}{(2\pi)^m}\cdot 2\pi\rho_1\ldots 2\pi\rho_m \cdot \frac{1}{\rho_1^2\ldots\rho_m^2}\cdot \sup_{\substack{|\lambda_j|=\rho_j\\ 1\leq j\leq m}} \|f(\xi+\sum_{j=1}^{m}\lambda_j x_j)\|$$

$$\leq \frac{1}{m!}\frac{1}{\rho_1\ldots\rho_m}\sup_{\|t-\xi\|\leq\rho}\|f(t)\| \leq \frac{1}{m!}\frac{M}{\rho_1\ldots\rho_m} < \infty.$$

Therefore $\|A_m\| = \sup_{\substack{\|x_j\|=1\\ 1\leq j\leq m}} \|A_m(x_1,\ldots,x_m)\| < \infty$, and so A_m is continuous.

Now let σ be a real number, $\sigma > 1$; then if $x \in B_{\rho\sigma^{-1}}(\xi)$, $(1-\lambda)\xi + \lambda x \in B_\rho(\xi) \subset U$ for every $\lambda \in \mathbb{C}$, $|\lambda| < \sigma$. We claim

$$\text{c)}\quad f(x) - \sum_{k=0}^{m} A_k(x-\xi)^k = \frac{1}{2\pi i}\int_{|\lambda|=\sigma} \frac{f[(1-\lambda)\xi+\lambda x]}{\lambda^{m+1}(\lambda-1)}\,d\lambda$$

for $f \in \overline{\mathcal{H}(U;F)}$, $x \in B_{\rho\sigma^{-1}}(\xi)$ and $m \in \mathbb{N}$.

We have already proved this for $f \in \mathcal{H}(U;F)$ (Proposition 4.5). If $\|x-\xi\| < \rho\sigma^{-1}$, the following mappings are continuous:

i) $V_x: h \in C(U;F) \mapsto V_x(h) = h(x) \in F$,

ii) $\psi_x: h \in C(U;F) \mapsto \frac{1}{2\pi i}\int_{|\lambda|=\sigma} \frac{h[(1-\lambda)\xi+\lambda x]}{\lambda^{m+1}(\lambda-1)}\,d\lambda.$

Therefore c) holds for $f \in \overline{\mathcal{H}(U;F)}$.

From c) we have

$$(*)\quad \|f(x)-\sum_{k=0}^{m} A_k(x-\xi)^k\| \leq \frac{1}{\sigma^m(\sigma-1)}\cdot\sup_{|\lambda|=\sigma}\|f[(1-\lambda)\xi+\lambda x]\|$$

for every $m \in \mathbb{N}$. Since $\|x-\xi\| \le \rho\sigma^{-1}$, we have for $\lambda \in \mathbb{C}$, $|\lambda| = \sigma$,

$$\|(1-\lambda)\xi + \lambda x - \xi\| = \sigma\|x-\xi\| < \rho,$$

and so, by 3), $\sup_{|\lambda|=\sigma} \|f[(1-\lambda)\xi+\lambda x]\| \le M$. Hence, by (*),

$$\|f(x) - \sum_{k=0}^{m} A_k(x-\xi)^k\| \le \frac{M}{\sigma^m(\sigma-1)} \text{ for every } m \in \mathbb{N}.$$

Since $\sigma > 1$, it follows that $\sum_{m=0}^{\infty} A_m(x-\xi)^m$ converges to $f(x)$ uniformly in $B_{\rho\sigma^{-1}}(\xi)$, and therefore $f \in \mathcal{H}(U;F)$.

Q.E.D.

CHAPTER 5

CONVERGENCE OF THE TAYLOR SERIES

In this chapter we consider the following problem: Given $f \in \mathcal{H}(U;F)$ and $\xi \in U$, we seek to determine the subsets of U in which the Taylor series of f at ξ converges uniformly to f; that is, we wish to know to what extent the Taylor series of f at ξ represents f. In the case in which $E = C^n$, it is well known that the Taylor series of f at ξ converges uniformly to f in every compact subset of the largest open ball centred at ξ and contained in U.

DEFINITION 5.1 If E is a normed space, $A \subset E$ and $\xi \in E$, A is said to be ξ-balanced if $(1-\lambda)\xi + \lambda x \in A$ for every $x \in A$ and $\lambda \in C$, $|\lambda| \leq 1$.

If A is ξ-balanced and non-empty, then $\xi \in A$. The open and closed balls with centre ξ are the simplest examples of ξ-balanced sets in a normed space E.

A subset A of E is ξ-balanced if and only if the set $A-\xi = \{x-\xi : x \in A\}$ is 0-balanced. The 0-balanced sets are referred to simply as balanced sets.

If $\xi \in A \subset E$, the set

$$A_\xi = \{x \in A : (1-\lambda)\xi + \lambda x \in A \text{ if } \lambda \in \mathbb{C}, |\lambda| \leq 1\}$$

is the largest ξ-balanced set contained in A; it is called the ξ-balanced kernel of A. A_ξ is never empty. If A is an open set then A_ξ is also open.

If $\xi \in A \subset E$ where A is ξ-balanced, and $B \subset A$, then

$$\tilde{B}_\xi = \{(1-\lambda)\xi+\lambda x : x \in B, \lambda \in \mathbb{C}, |\lambda| \leq 1\} \subset A.$$

Thus $\tilde{B}_\xi$ is the smallest ξ-balanced set containing B. $\tilde{B}_\xi$ is called the ξ-balanced hull of B.

PROPOSITION 5.1 Let U be an open ξ-balanced set and let $f \in \mathcal{H}(U;F)$. For every compact set K, $K \subset U$, there exists an open set V, $K \subset V \subset U$, and a number $\rho > 1$, such that

$$\{(1-\lambda)\xi+\lambda x : x \in V, \lambda \in \mathbb{C}, |\lambda| \leq \rho\} \subset U$$

and

$$\sup\{\|f[(1-\lambda)\xi+\lambda x]\| : x \in V, \lambda \in \mathbb{C}, |\lambda| \leq \rho\} < \infty.$$

PROOF. Denoting by $\bar{D}_\rho$ the closed ball in $\mathbb{C}$ with centre 0 and radius ρ, the mapping

$$T: (\lambda,x) \in \mathbb{C}\times E \mapsto (1-\lambda)\xi + \lambda x \in E$$

is continuous, and so $T(\bar{D}_1 \times K)$ is a compact set contained in U. Since $f \in \mathcal{H}(U;F) \subset C(U;F)$, $f[T(\bar{D}_1\times K)]$ is a compact subset of F, and hence is bounded. Therefore there exist a bounded neighbourhood A of $f[T(\bar{D}_1\times K)]$ in F, a real number $\rho > 1$, and an open subset V of E, $K \subset V \subset U$, such that

$$\bar{D}_\rho \times V \subset T^{-1}\{f^{-1}(A)\}$$

which implies that

$$T(\bar{D}_\rho \times V) \subset f^{-1}(A) \quad \text{and} \quad f[T(\bar{D}_\rho \times V)] \subset A.$$

Since $f^{-1}(A) \subset U$, we have

$$(1-\lambda)\xi + \lambda x \in U \quad \text{if} \quad x \in V \quad \text{and} \quad \lambda \in \mathbb{C}, \ |\lambda| \le \rho,$$

and

$$\sup\{\|f[(1-\lambda)\xi+\lambda x]\| : x \in V, \lambda \in \mathbb{C}, |\lambda| \le \rho\} < \infty.$$

Q.E.D.

REMARK 5.1: Proposition 5.1 can also be stated in the following way: "Let $\xi \in U$, K a compact ξ-balanced subset of U, and $f \in \mathcal{H}(U;F)$. Then for every open set V, $K \subset V \subset U$, there exists an open ξ-balanced set W, $K \subset W \subset V$, and a real number $\rho > 1$ such that

$$\{(1-\lambda)\xi+\lambda x : x \in W, \lambda \in \mathbb{C}, |\lambda| \le \rho\} \subset U$$

and

$$\sup\{\|f[(1-\lambda)\xi+\lambda x]\| : x \in W, \lambda \in \mathbb{C}, |\lambda| \le \rho\} < \infty."$$

PROPOSITION 5.2 Let $f \in \mathcal{H}(U;F)$ and $\xi \in U$, where U is an open, ξ-balanced set. Then for every compact set K, $K \subset U$, there is a neighbourhood V of K, $V \subset U$, such that the Taylor series of f at ξ converges uniformly to f in V.

PROOF. Given a compact set $K \subset U$, by Proposition 5.1 there exists a real number $\rho > 1$ and an open set V such that $K \subset V \subset U$ and

$$\sup\{\|f[(1-\lambda)\xi+\lambda x]\| : x \in V, \lambda \in \mathbb{C}, |\lambda| \le \rho\} = L < \infty.$$

By Corollary 4.1 we have

$$\|f(x) - \sum_{m=0}^{\ell} \frac{1}{m!} \hat{d}^m f(\xi)(x-\xi)\| \leq \frac{L}{\rho^{\ell}(\rho-1)}$$

for every $\ell \geq 0$ and every $x \in V$. Since $\rho > 1$, the proposition follows immediately. Q.E.D.

COROLLARY 5.1 Let $f \in \mathcal{H}(U;F)$ and $\xi \in U$, where U is an open ξ-balanced set. Then the Taylor series of f at ξ converges to f at every point of U.

COROLLARY 5.2 Let $f \in \mathcal{H}(U;F)$ and $\xi \in U$. Then the Taylor series of f at ξ converges uniformly to f in a neighbourhood contained in U of every compact set contained in U_ξ.

REMARK 5.2: Corollary 5.2 shows that for convergence of the Taylor series, the largest open ball with centre ξ contained in U of the finite dimensional case is replaced in the infinite dimensional case by the ξ-balanced set U_ξ.

DEFINITION 5.2 A mapping $f: U \to F$ is said to be locally bounded at $\xi \in U$ if there exists a neighbourhood V of ξ, $V \subset U$, such that f is bounded in V; f is said to be locally bounded in U if it is locally bounded at every point of U.

DEFINITION 5.3 Let $f: U \to F$ be locally bounded at $\xi \in U$. The radius of boundedness of f at ξ is the supremum of the set of real numbers $r > 0$ such that $B_r(\xi) \subset U$ and f is bounded in every ball $\bar{B}_\rho(\xi)$ with $0 < \rho < r$.

PROPOSITION 5.3 Let F be a Banach space, $f \in \mathcal{H}(U;F)$ and $\xi \in U$. The radius of boundedness r_b of f at ξ is equal to the minimum of the radius of convergence R of the Taylor series of f at ξ and the distance d from ξ to the boundary of U.

PROOF. From Definition 5.3 we have $r_b \leq d$, and also, if ρ is a real number, $0 < \rho < r_b$, then $L = \sup_{\|x-\xi\|=\rho} \|f(x)\|$ is finite. Applying the Cauchy inequalities, we have

$$\|P_m\| = \|\frac{1}{m!} \hat{d}^m f(\xi)\| \leq \frac{L}{\rho^m}$$

for every $m \in \mathbb{N}$. Therefore $\overline{\lim} \|P_m\|^{1/m} \leq \frac{1}{\rho}$, and so by the Cauchy-Hadamard formula

$R = (\overline{\lim}\|P_m\|^{1/m})^{-1} \geq \rho$, where $\overline{\lim}\|P_m\|^{1/m}$ denotes $\lim\sup_{m\to\infty} \|P_m\|^{1/m}$

Hence $r_b \leq R$, and therefore

$$(*) \qquad r_b \leq \min(d,R).$$

To prove the reverse of inequality (*), let $0 < \rho < \min(d,R)$. Since $\rho < R$, by the Cauchy-Hadamard formula the Taylor series of f at ξ converges uniformly in $\bar{B}_\rho(\xi)$, and since $\bar{B}_\rho(\xi)$ is ξ-balanced, by Corollary 5.1 this series converges to f at every point of $\bar{B}_\rho(\xi)$. Therefore the Taylor series of f at ξ converges uniformly to f in $\bar{B}_\rho(\xi)$. Since $P_m \in \mathcal{P}(^mE;F)$, $m \in \mathbb{N}$, P_m is bounded in $\bar{B}_\rho(0)$. It follows that f is bounded in $\bar{B}_\rho(\xi)$, and therefore

$$(**) \qquad r_b \geq \min(d,R).$$

In the case in which $\partial U = \emptyset$, we set $d = \infty$. Q.E.D.

REMARK 5.3: In the classical case in which $E = \mathbb{C}^n$, $n \geq 1$, given $\xi \in U$, if $\rho \in \mathbb{R}$, $0 < \rho < \text{dist}(\xi, \partial U)$ then $\bar{B}_\rho(\xi) \subset U$ and $\bar{B}_\rho(\xi)$ is compact. Therefore

$$\sup_{\|x-\xi\| \leq \rho} \|f(x)\| < \infty$$

for every $f \in \mathcal{H}(U;F) \subset C(U;F)$. From the definition of the radius of boundedness and Proposition 5.3 it follows that $r_b = d \leq R$; this explains why, in the finite dimensional case, the concept of radius of boundedness is not of great significance. However, in the case in which the dimension of E is infinite, it is possible to have $r_b = R < d$.

EXAMPLE 5.1 Let E be C_{oo}, the space of all sequences $x = (x_1, \ldots, x_m, \ldots)$ of complex numbers which are eventually zero. Thus for each $x \in C_{oo}$ there is an index m_o, depending on x, such that $x_{m_o} = x_{m_o+1} = \ldots = 0$. The norm of this space is given by

$$x = (x_1, \ldots, x_m, \ldots) \mapsto \|x\| = \sup_m |x_m|.$$

This normed space is not a Banach space. Taking $F = \mathbb{C}$, we define $P_m : E \to F$ by $P_m(x) = x_1 \ldots x_m$ if $x = (x_1, \ldots, x_m, \ldots) \in C_{oo}$ and $m \in \mathbb{N}$, $m \geq 1$, and $P_o(x) = 0$. It is easy to see that P_m is a continuous m-homogeneous polynomial from E into F, and that $\|P_m\| = 1$, $m \in \mathbb{N}$, $m \geq 1$. Consider the power series $\sum_{m=0}^{\infty} P_m(x)$. We claim:

a) This series converges for every $x \in E$;

b) If $0 < \rho < 1$ the series converges uniformly in $\bar{B}_\rho(t)$ for every $t \in E$;

c) The mapping $f: E \to F$ defined by $f(x) = \sum_{m=0}^{\infty} P_m(x)$, $x \in E$, is holomorphic in E, that is, f is entire.

For every $x \in E$, $P_m(x) = 0$ after a certain index; this proves a).

Let $t = (t_1,\ldots,t_m,\ldots) \in E$ and $x \in \bar{B}_\rho(t)$, where $0 < \rho < 1$. Then $x = t+y$, $\|y\| \le \rho$, and if m_o is an index such that $t_m = 0$ for $m > m_o$, we have, with $k = 0,1,2,\ldots$:

$$(|y_1|+|t_1|)\ldots(|y_{m_o+1}|+|t_{m_o+1}|)\ldots(|y_{m_o+k}|+(t_{m_o+k}|) \le$$
$$C|y_{m_o+1}|\ldots|y_{m_o+k}| \le C\cdot\rho^k ,$$

where

$$C = (|t_1|+\rho)\ldots(|t_{m_o}|+\rho).$$

Hence $\sum_{m=m_o}^{\infty} |P_m(x)| \le C + C\cdot\rho + C\cdot\rho^2 + \ldots$, which shows that the series $\sum_{m=0}^{\infty} P_m(x)$ converges uniformly in $\bar{B}_\rho(t)$. This proves b).

From this it follows, using a classical compactness argument, that $f \in C(E;F)$; since $\mathcal{H}(E;F)$ is closed in $C(E;F)$ for the compact-open topology, we then have $f \in \mathcal{H}(E;F)$, and c) is proved.

In this example, with $\xi = 0$, we have $d = \infty$, and by the Cauchy-Hadamard formula, $R = 1$. Therefore $r_b = \min(\infty,1) = 1$.

REMARK 5.4: In Example 5.1 we have a phenomenon which cannot occur in the finite dimensional case - a holomorphic, and consequently continuous, function which is not bounded on every bounded subset of its domain. We note that it can be proved directly that f is not bounded in $\bar{B}_1(0)$: for each $m \in \mathbb{N}$, $m \geq 1$, consider the point $x^m = (\overbrace{1,\ldots,1}^{m},0,0,\ldots) \in \partial\bar{B}_1(0)$. We have

$$P_o(x^m) = 0,\ P_1(x^m) = 1,\ldots,P_m(x^m) = 1,\ P_{m+1}(x^m) = 0,\ldots ,$$

and so $f(x^m) = m$ for $m \in \mathbb{N}$, $m \geq 1$. Therefore $\sup_{\|x\| \leq 1} \|f(x)\| = \infty$.

EXAMPLE 5.2 Let c_o be the vector space of all sequences $x = (x_1,\ldots,x_m,\ldots)$ of complex numbers such that $\lim_{m\to\infty} x_m = 0$, with the norm

$$x \in c_o \mapsto \|x\| = \sup_m |x_m|.$$

c_o is a Banach space. Taking $F = \mathbb{C}$, we define $P_o(x) = 0$, and $P_m(x) = (x_m)^m$ if $m \in \mathbb{N}$, $m \geq 1$, and $x = (x_1,\ldots,x_m,\ldots) \in E$. Then P_m is a continuous m-homogeneous polynomial from E into F for every $m \in N$. Consider the power series $\sum_{m=0}^{\infty} P_m(x)$. We have:

a) The series converges for every $x \in E$;

b) If $0 < \rho < 1$ the series converges uniformly in $\bar{B}_\rho(t)$ for every $t \in E$;

c) The mapping $f: E \to F$, defined by $f(x) = \sum_{m=0}^{\infty} P_m(x)$, $x \in E$, is holomorphic in E.

Taking $\xi = 0$ we again have $r_b = R = 1 < d$. Thus we have another example of an entire function which is not bounded on every bounded subset of its domain. Note that c_o is separable but not reflexive, since $c_o' = \ell^1$ and $(\ell^1)'$ is the space of bounded sequences. However, this phenomenon can occur even when E is separable and reflexive.

EXAMPLE 5.3 Let $E = \ell^p$, $p \in \mathbb{N}$, $p > 1$, and $F = \mathbb{C}$. Using the same polynomials and the same function as in Example 5.2, we again find that $r_b = R = 1 < d$ at $\xi = 0$.

REMARK 5.5: We call the attention of the reader to Chapters 11, 12 and 13.

PROPOSITION 5.4 If $f \in \mathcal{H}(U;F)$ and $m \in \mathbb{N}$ then $d^m f \in \mathcal{H}(U;\mathcal{L}_s(^mE;F))$ and $\hat{d}^m f \in \mathcal{H}(U;\mathcal{P}(^mE;F))$. If $f(x) \simeq \simeq \sum_{m=0}^{\infty} P_m(x-\xi)$ is the Taylor series of f at $\xi \in U$, then the Taylor series of $d^m f$ and $\hat{d}^m f$ at ξ are

$$d^m f(x) \simeq \sum_{k=0}^{\infty} d^m P_{k+m}(x-\xi)$$

$$\hat{d}^m f(x) \simeq \sum_{k=0}^{\infty} \hat{d}^m P_{k+m}(x-\xi).$$

PROOF. Let $g_\lambda : U \to F$, $\lambda \in \mathbb{N}$, be defined by

$$g_\lambda(x) = f(x) - \sum_{i=0}^{m+\lambda} P_i(x-\xi), \quad x \in U;$$

$t_\xi : x \in E \mapsto x-\xi \in E$, and $Q_i = P_i \circ t_\xi$. By Propositions 1.6 and 3.1 we have $Q_i \in \mathcal{P}(E;F) \subset \mathcal{H}(E;F)$ for every $i \in \mathbb{N}$. Using the same notation, Q_i, for the restriction of this polynomial to U, $Q_i \in \mathcal{H}(U;F)$ for every $i \in \mathbb{N}$, and since $g_\lambda = f - \sum_{i=0}^{m+\lambda} Q_i$, $g_\lambda \in \mathcal{H}(U;F)$ for every $\lambda \in \mathbb{N}$.

Since $\hat{d}^m : \mathcal{H}(U;F) \to \mathcal{F}(U;\mathcal{P}(^mE;F))$ is linear we have

$$\hat{d}^m g_\lambda = \hat{d}^m f - \sum_{i=0}^{m+\lambda} \hat{d}^m Q_i ,$$

and therefore

$$(*) \qquad \hat{d}^m g_\lambda(x) = \hat{d}^m f(x) - \sum_{i=0}^{m+\lambda} \hat{d}^m P_i(x-\xi)$$

for every $x \in U$, $\lambda \in \mathbb{N}$.

From (*) and Proposition 3.1 we have

$$\hat{d}^m g_\lambda(x) = \hat{d}^m f(x) - \sum_{i=m}^{m+\lambda} \hat{d}^m P_i(x-\xi)$$
$$= \hat{d}^m f(x) - \sum_{k=0}^{\lambda} \hat{d}^m P_{k+m}(x-\xi)$$

for every $x \in U$, $\lambda \in \mathbb{N}$.

Let r_b be the radius of boundedness of f at ξ, let ρ and ρ' be real numbers such that $0 < \rho < \rho' < r_b$, $\sigma = \rho' - \rho$, and $X = \{x \in U : \bar{B}_\sigma(x) \subset U\}$. Then, by the Cauchy inequalities, $\|\hat{d}^m g_\lambda(x)\| \le \frac{m!}{\sigma^m} \cdot \sup_{\|t-x\|=\sigma} \|g_\lambda(t)\|$ for every $x \in X$, and hence

$$(**) \qquad \|\hat{d}^m f(x) - \sum_{k=0}^{\lambda} \hat{d}^m P_{k+m}(x-\xi)\| \le$$

$$\frac{m!}{\sigma^m} \cdot \sup_{\|t-x\|=\sigma} \|f(t) - \sum_{i=0}^{m+\lambda} P_i(t-\xi)\| \quad \text{for every } x \in X.$$

Since $\rho' < r_b \le \operatorname{dist}(\xi, \partial U)$, $\bar{B}_\sigma(x) \subset \bar{B}_{\rho'}(\xi) \subset U$ for every $x \in \bar{B}_\rho(\xi)$, which implies that $\bar{B}_\rho(\xi) \subset X$. Furthermore, if $x \in \bar{B}_\rho(\xi)$ and $\|t-x\| = \sigma$ then $\|t-\xi\| \le \rho'$; therefore, by (**) we have

$$\|\hat{d}^m f(x) - \sum_{k=0}^{\lambda} \hat{d}^m P_{k+m}(x-\xi)\| \leq$$

$$\frac{m!}{\sigma^m} \cdot \sup_{\|t-x\| \leq \rho'} \|f(t) - \sum_{i=0}^{m+\lambda} P_i(t-\xi)\|$$

for every $x \in \bar{B}_\rho(\xi)$. Since $\rho' < r_b$, the second term of this inequality tends to 0 as λ tends to ∞, which shows that $\sum_{k=0}^{\infty} \hat{d}^m P_{k+m}(x-\xi)$ converges uniformly to $\hat{d}^m f(x)$ in $\bar{B}_\rho(\xi)$.

This proves the statement in the proposition concerning $\hat{d}^m f$, since $\hat{d}^m P_{k+m} \in \mathcal{P}({}^k E; \mathcal{P}({}^m E;F))$ for every $k \in \mathbb{N}$ (Corollary 3.1). To prove the corresponding statements for $d^m f$, it suffices to note that the equation

$$\hat{d}^m f(x) - \sum_{k=0}^{M} \hat{d}^m P_{k+m}(x-\xi) = \sum_{k>M} \hat{d}^m P_{k+m}(x-\xi)$$

can be written as

$$\widehat{d^m f(x) - \sum_{k=0}^{M} d^m P_{k+m}(x-\xi)} = \sum_{k>M} \hat{d}^m P_{k+m}(x-\xi)$$

and, by Proposition 1.3,

$$\|d^m f(x) - \sum_{k=0}^{M} d^m P_{k+m}(x-\xi)\| \leq \frac{m^m}{m!} \|\widehat{d^m f(x) - \sum_{k=0}^{M} d^m P_{k+m}(x-\xi)}\|.$$

Q.E.D.

COROLLARY 5.3 If $f \in \mathcal{H}(U;F)$ and $\xi \in U$, then for $k,m \in \mathbb{N}$,

$$\frac{1}{k!} \hat{d}^k[\frac{1}{m!} \hat{d}^m f](\xi) = \frac{1}{m!} \hat{d}^m[\frac{1}{(k+m)!} \hat{d}^{k+m} f(\xi)].$$

PROOF. In the notation of the proposition we have

$$\frac{1}{k!} \hat{d}^k(\hat{d}^m f)(\xi) = \hat{d}^m P_{k+m} = \hat{d}^m[\frac{1}{(k+m)!} \hat{d}^{k+m} f(\xi)].$$

Q.E.D.

COROLLARY 5.4 If $f \in \mathcal{H}(U;F)$ and $\xi \in U$, then for $k,m \in \mathbb{N}$,

$$\tau_{k,\hat{d}^m f,\xi} = \hat{d}^m[\tau_{k+m,f,\xi}].$$

CHAPTER 6

WEAK HOLOMORPHY

In this chpater we shall prove the following propositions:

PROPOSITION 6.2 If F is a Banach space, and $f: U \to F$ a mapping, the following are equivalent:

1) $f \in \mathcal{H}(U;F)$.

2) $\psi \circ f \in \mathcal{H}(U;\mathbb{C})$ for every $\psi \in F'$, where F' is the topological dual of F.

PROPOSITION 6.3 Let E, F and G be normed spaces, F and G complete, U a non-empty open subset of E, and $f: U \to \mathcal{L}(F;G)$ a mapping. Then the following are equivalent:

a) $f \in \mathcal{H}(U;\mathcal{L}(F;G))$.

b) The mapping $x \in U \mapsto \omega[f(x)(y)] \in \mathbb{C}$ is holomorphic in U for every $y \in F$ and $\omega \in G'$.

PROPOSITION 6.4 If F is a Banach space and $f: U \to F'$ a mapping, the following are equivalent:

a) $f \in \mathcal{H}(U;F')$

b) The mapping $x \in U \mapsto f(x)(y) \in \mathbb{C}$ is holomorphic in U for every $y \in F$.

We begin with:

LEMMA 6.1 Let M be a metric space, F a normed space (real or complex) and $f: M \to F$ a mapping. The following are equivalent:

1) f is locally bounded (Definition 5.2).

2) f is bounded on every compact subset of M.

PROOF:

1) $\Rightarrow$ 2). This is proved by a simple compactness argument. We note that this implication is valid if M is an arbitrary topological space.

2) $\Rightarrow$ 1). Let $\xi \in M$, and suppose that f is not bounded on any neighbourhood of ξ. Then, by recursion, we obtain for each $m \in \mathbb{N}$, $m \geq 1$, a point $x_m \in M$ such that

a) $d(\xi, x_m) < \frac{1}{m}$ and b) $\|f(x_m)\| > m$.

By a), $x_m \to \xi$ as $m \to \infty$, and hence the set $K = \{\xi\} \cup \cup \{x_m; m \geq 1\}$ is compact in M. But by b), f is unbounded on K, which is a contradiction. Q.E.D.

PROPOSITION 6.1 Let F be a Banach space and $f: U \to F$ a mapping. Suppose that $\Psi \subset F'$ is such that $Y \subset F$ is bounded if and only if $\psi(Y)$ is bounded for every $\psi \in \Psi$. Then the following are equivalent:

1) $f \in \mathcal{H}(U;F)$

2) $\psi \circ f \in \mathcal{H}(U,\mathbb{C})$ for every $\psi \in \Psi$.

PROOF:

1) $\Rightarrow$ 2). This implication is true without any restrictions on Ψ, by virtue of Proposition 3.2.

2) $\Rightarrow$ 1)

a) Ψ separates the points of F, that is, for every $y \in F$, $y \neq 0$, there exists $\psi \in \Psi$ such that $\psi(y) \neq 0$. For if $y \neq 0$, the set $\mathbb{C}y = \{\lambda y : \lambda \in \mathbb{C}\}$ is unbounded in F, and so by the definition of Ψ there exists $\psi \in \Psi$ such that $\psi(\mathbb{C}y) = \mathbb{C}\psi(y)$ is unbounded; hence $\psi(y) \neq 0$.

b) For every real number $d \geq 0$ there exists a real number $c \geq 0$ such that $\|y\| \leq cd$ for every $y \in F$ such that $|\psi(y)| \leq d\|\psi\|$ for all $\psi \in \Psi$. In fact, let $Y = \{y \in F : |\psi(y)| \leq d\|\psi\|$ for all $\psi \in \Psi\}$. Y is non-empty, since $0 \in Y$, and if $\psi \in \Psi$ we have $\sup_{y \in Y} |\psi(y)| \leq d\|\psi\|$, which implies, by the definition of Ψ, that Y is bounded in F. Taking $c = d^{-1} \sup_{y \in Y} \|y\|$ if $d \neq 0$, it follows that $\|y\| \leq cd$ for every $y \in Y$. If $d = 0$, the assertion is trivial.

c) f is bounded on every compact $K \subset U$: Since $\psi \circ f$ is holomorphic, and hence continuous, for every $\psi \in \Psi$, $\psi \circ f$ is bounded on K for every $\psi \in \Psi$. Thus $\psi[f(K)]$ is bounded for every $\psi \in \Psi$, and so, by the definition of Ψ, $f(K)$ is a bounded subset of F.

d) f is locally bounded: This follows from c) and Lemma 6.1, since U is a metric space.

e) f is continuous in U: Let $\xi \in U$. Since U is open, there exists $r > 0$ such that $\bar{B}_r(\xi) \subset U$. Applying Corollary 4.2 to $\psi \circ f \in \mathcal{H}(U;\mathbb{C})$ with $m = 0$, $\psi \in \Psi$, we have

$$|\psi\circ f(x) - \psi\circ f(\xi)| \le \frac{\|x-\xi\|}{r-\|x-\xi\|} \sup_{\|t-\xi\|=r} |\psi\circ f(t)|$$

$$\le \frac{\|x-\xi\|}{r-\|x-\xi\|} \|\psi\| \sup_{\|t-\xi\|=r} \|f(t)\|$$

for every $x \in B_r(\xi)$ and $\psi \in \Psi$.

By d) f is locally bounded, and so, taking r sufficiently small, so that $\sup_{\|t-\xi\|=r} \|f(t)\| = d < +\infty$, we have

$$|\psi(f(x)-f(\xi))| \le \frac{\|x-\xi\|}{r-\|x-\xi\|} \|\psi\| d$$

for every $x \in B_r(\xi)$ and $\psi \in \Psi$. Therefore

$$\left|\psi\left[\frac{r-\|x-\xi\|}{\|x-\xi\|} (f(x)-f(\xi))\right]\right| \le d\|\psi\|$$

for every $x \in B_r(\xi)$, $x \neq \xi$, and $\psi \in \Psi$. It follows by b) that there exists a real number $c \ge 0$ such that

$$\left\|\frac{r-\|x-\xi\|}{\|x-\xi\|} (f(x)-f(\xi))\right\| \le cd$$

for every $x \in B_r(\xi)$, $x \neq \xi$, and hence

$$\|f(x)-f(\xi)\| \le \frac{\|x-\xi\|}{r-\|x-\xi\|} cd.$$

It follows that f is continuous at ξ.

f) f is holomorphic in U: Let $\xi \in U$. For each $m \in \mathbb{N}$ we define a mapping $P_m: E \to F$ in the following way: For each $x \in E$ we choose a real number $\rho > 0$ such that $\xi+\lambda x \in U$ for every $\lambda \in \mathbb{C}$, $|\lambda| \le \rho$, and set

$$P_m(x) = \frac{1}{2\pi i} \int_{|\lambda|=\rho} \frac{f(\xi+\lambda x)}{\lambda^{m+1}} d\lambda .$$

This integral exists since by e) f is continuous, and by hypothesis F is complete. We show first that the value of the integral is independent of ρ. Let ρ, ρ' be positive real numbers such that $\xi+\lambda x \in U$ for $\lambda \in \mathbb{C}$, $|\lambda| \leq \max\{\rho,\rho'\}$, and let

$$P_m(x) = \frac{1}{2\pi i} \int_{|\lambda|=\rho} \frac{f(\xi+\lambda x)}{\lambda^{m+1}} d\lambda ;$$

$$P'_m(x) = \frac{1}{2\pi i} \int_{|\lambda|=\rho'} \frac{f(\xi+\lambda x)}{\lambda^{m+1}} d\lambda .$$

By hypothesis, $\psi \circ f \in \mathcal{H}(U;\mathbb{C})$ for every $\psi \in \Psi$, and hence, by Proposition 4.2 (the Cauchy integral formula),

$$\frac{1}{m!} \hat{d}^m(\psi\circ f)(\xi)(x) = \frac{1}{2\pi i} \int_{|\lambda|=\rho} \frac{(\psi\circ f)(\xi+\lambda x)}{\lambda^{m+1}} d\lambda$$

$$= \psi\Big[\frac{1}{2\pi i} \int_{|\lambda|=\rho} \frac{f(\xi+\lambda x)}{\lambda^{m+1}} d\lambda\Big] = \psi[P_m(x)] ,$$

for every $\psi \in \Psi$ and $m \in \mathbb{N}$. Similarly,

$$\frac{1}{m!} \hat{d}^m(\psi\circ f)(\xi)(x) = \psi[P'_m(x)]$$

for every $\psi \in \Psi$ and $m \in \mathbb{N}$. Therefore $\psi[P_m(x)] = \psi[P'_m(x)]$ for every $\psi \in \Psi$ and $m \in \mathbb{N}$. By a) Ψ separates the points of F, and hence $P_m(x) = P'_m(x)$ for every $m \in \mathbb{N}$.

We show next that the mappings P_m, $m \in \mathbb{N}$, are continuous. If $x_o \in E$, the mapping $(\lambda, x) \in \mathbb{C}\times E \to \xi+\lambda x \in E$ is continuous at the point $(0,x_o)$, and so there exists $\rho > 0$ such that $\xi+\lambda x \in U$ for all $\lambda \in \mathbb{C}$ and $x \in E$ with $|\lambda| \leq \rho$ and $\|x-x_o\| \leq \rho$. Then

$$P_m(x) = \frac{1}{2\pi i}\int_{|\lambda|=\rho} \frac{f(\xi+\lambda x)}{\lambda^{m+1}}\, d\lambda$$

for every $x \in B_\rho(x_o)$. Now, using the fact that f is continuous (part e)), and the usual inequalities for integrals, it follows that P_m is continuous at x_o for every $m \in \mathbb{N}$.

We now show that $P_m \in \mathcal{P}(^mE;F)$ for every $m \in \mathbb{N}$. For $m = 0$ this is immediate, since

$$P_o(x) = \frac{1}{2\pi i}\int_{|\lambda|=\rho} \frac{f(\xi+\lambda x)}{\lambda}\, d\lambda,$$

and applying Proposition 4.2 (the Cauchy integral formula) to the functions $\psi\circ f \in \mathcal{H}(U;\mathbb{C})$, $\psi \in \Psi$,

$$\psi[P_o(x)] = \frac{1}{2\pi i}\int_{|\lambda|=\rho} \frac{(\psi\circ f)(\xi+\lambda x)}{\lambda}\, d\lambda =$$

$$= \frac{1}{0!}\hat{d}^o(\psi\circ f)(\xi)(x) = \psi[f(\xi)]$$

for every $x \in E$ and $\psi \in \Psi$. Since, by a), Ψ separates the points of F, $P_o(x) = f(\xi)$ for every $x \in E$. Therefore $P_o \in \mathcal{P}(^oE;F)$.

For $m \in \mathbb{N}$, $m \geq 1$, we define a mapping $A_m\colon E^m \to F$ by

$$A_m(x_1,\ldots,x_m) = \frac{1}{2^m m!} \sum_{\substack{\varepsilon_i=\pm 1\\ 1\le i\le m}} \varepsilon_1\varepsilon_2\cdots\varepsilon_m\, P_m(\varepsilon_1 x_1+\ldots+\varepsilon_m x_m)$$

where $(x_1,\ldots,x_m) \in E^m$. We shall prove that the mappings A_m, $m \in \mathbb{N}$, $m \ge 1$, are symmetric and m-linear, and that $\hat{A}_m = P_m$. It is easy to see that A_m is symmetric, and it then suffices to prove that A_m is linear in the first variable. We show only that $A_m(x_1+x_1',x_2,\ldots,x_m) = A_m(x_1,x_2,\ldots,x_m) + A_m(x_1',x_2,\ldots,x_m)$ for every $x_1,x_1',x_2,\ldots,x_m \in E$.

Since $\psi\circ P_m = \frac{1}{m!}\hat{d}^m(\psi\circ f)(\xi)$ for every $m \in \mathbb{N}$ and $\psi \in \Psi$, we have

$$\psi[A_m(x_1+x_1',x_2,\ldots,x_m)] =$$

$$\frac{1}{2^m m!} \sum_{\substack{\varepsilon_i=\pm 1\\ 1\le i\le m}} \varepsilon_1\cdots\varepsilon_m(\psi\circ P_m)(\varepsilon_1(x_1+x_1')+\varepsilon_2 x_2+\ldots+\varepsilon_m x_m) =$$

$$= \frac{1}{2^m m!} \sum_{\substack{\varepsilon_i=\pm 1\\ 1\le i\le m}} \varepsilon_1\cdots\varepsilon_m \frac{1}{m!}\hat{d}^m(\psi\circ f)(\xi)(\varepsilon_1(x_1+x_1')+\varepsilon_2 x_2+\ldots+\varepsilon_m x_m) =$$

$$= \frac{1}{m!} d^m(\psi\circ f)(\xi)(x_1+x_1',x_2,\ldots,x_m) =$$

$$= \frac{1}{m!} d^m(\psi\circ f)(\xi)(x_1,x_2,\ldots,x_m) + \frac{1}{m!} d^m(\psi\circ f)(\xi)(x_1',x_2,\ldots,x_m) =$$

$$= \frac{1}{2^m m!} \sum_{\substack{\varepsilon_i=\pm 1\\ 1\le i\le m}} \varepsilon_1\cdots\varepsilon_m \frac{1}{m!}\hat{d}^m(\psi\circ f)(\xi)(\varepsilon_1 x_1+\ldots+\varepsilon_m x_m) +$$

$$\frac{1}{2^m m!} \sum_{\substack{\varepsilon_i=\pm 1\\ 1\le i\le m}} \varepsilon_1\cdots\varepsilon_m \frac{1}{m!}\hat{d}^m(\psi\circ f)(\xi)(\varepsilon_1 x_1'+\ldots+\varepsilon_m x_m)$$

$$= \frac{1}{2^m m!} \sum_{\substack{\varepsilon_i=\pm 1\\ 1\le i\le m}} \varepsilon_1\cdots\varepsilon_m(\psi\circ P_m)(\varepsilon_1 x_1+\ldots+\varepsilon_m x_m) +$$

$$\frac{1}{2^m m!} \sum_{\substack{\epsilon_i=\pm 1 \\ 1\le i\le m}} \epsilon_1\cdots\epsilon_m(\psi\circ P_m)(\epsilon_1 x_1'+\cdots+\epsilon_m x_m)$$

$$= \psi[\frac{1}{2^m m!} \sum_{\substack{\epsilon_i=\pm 1 \\ 1\le i\le m}} \epsilon_1\cdots\epsilon_m \, P_m(\epsilon_1 x_1+\cdots+\epsilon_m x_m) +$$

$$\frac{1}{2^m m!} \sum_{\substack{\epsilon_i=\pm 1 \\ 1\le i\le m}} \epsilon_1\cdots\epsilon_m \, P_m(\epsilon_1 x_1'+\cdots+\epsilon_m x_m)]$$

$$= \psi[A_m(x_1,x_2,\ldots,x_m) + A_m(x_1',x_2,\ldots,x_m)].$$

Therefore $\psi[A_m(x_1+x_1',x_2,\ldots,x_m)] =$
$= \psi[A_m(x_1,x_2,\ldots,x_m) + A_m(x_1',x_2,\ldots,x_m)]$ for every $\psi \in \Psi$. Since Ψ separates the points of F, it follows that

$$A_m(x_1+x_1',x_2,\ldots,x_m) = A_m(x_1,x_2,\ldots,x_m) + A_m(x_1',x_2,\ldots,x_m).$$

Next, we show that $\hat{A}_m = P_m$, from which it follows that $P_m \in \mathcal{P}({}^mE;F)$. We have

$$\psi[\hat{A}_m(x)] = \psi[A_m(\overbrace{x,\ldots,x}^{m})]$$

$$= \frac{1}{2^m m!} \sum_{\substack{\epsilon_i=\pm 1 \\ 1\le i\le m}} \epsilon_1\cdots\epsilon_m(\psi\circ P_m)((\epsilon_1+\cdots+\epsilon_m)x)$$

$$= \frac{1}{2^m m!} \sum_{\substack{\epsilon_i=\pm 1 \\ 1\le i\le m}} \epsilon_1\cdots\epsilon_m \frac{1}{m!}\hat{d}^m(\psi\circ f)(\xi)((\epsilon_1+\cdots+\epsilon_m)x)$$

$$= \frac{1}{m!} d^m(\psi\circ f)(\xi)(\overbrace{x,\ldots,x}^{m}) = \frac{1}{m!}\hat{d}^m(\psi\circ f)(\xi)(x)$$

$$= \psi[P_m(x)]$$

for every $\psi \in \Psi$, $x \in E$ and $m \in \mathbb{N}$, $m \ge 1$, and hence

$\hat{A}_m(x) = P_m(x)$ for every $x \in E$ and $m \in \mathbb{N}$, $m \geq 1$.

To complete the proof of the proposition we show that $\sum_{m=0}^{\infty} P_m(x-\xi)$ converges uniformly to f in a neighbourhood of ξ.

Since f is locally bounded (part d)) there exist real numbers $M > 0$ and $\sigma > 0$ such that $\bar{B}_\sigma(\xi) \subset U$ and $\sup_{\|t-\xi\|=\sigma} \|f(t)\| \leq M$. Choose real numbers $\rho > 1$ and $r > 0$ such that $r < \sigma\rho^{-1}$. We claim that the series $\sum_{m=0}^{\infty} P_m(x-\xi)$ converges uniformly to f in $B_r(\xi)$.

For every $x \in B_r(\xi)$ we have $(1-\lambda)\xi + \lambda x \in \bar{B}_\sigma(\xi) \subset U$ for all $\lambda \in \mathbb{C}$, $|\lambda| \leq \rho$, and so, applying Proposition 4.5 to the functions $\psi \circ f$, $\psi \in \Psi$,

$$\psi \circ f(x) - \sum_{k=0}^{m} \frac{1}{k!} \hat{d}^k(\psi \circ f)(\xi)(x-\xi) = \frac{1}{2\pi i} \int_{|\lambda|=\rho} \frac{\psi \circ f[(1-\lambda)\xi + \lambda x]}{\lambda^{m+1}(\lambda - 1)} d\lambda$$

for every $x \in B_r(\xi)$ and $m \in \mathbb{N}$. Using the fact that $\psi[P_k(x)] = \frac{1}{k!} \hat{d}^k(\psi \circ f)(\xi)(x)$, this yields

$$\psi[f(x) - \sum_{k=0}^{m} P_k(x-\xi)] = \psi[\frac{1}{2\pi i} \int_{|\lambda|=\rho} \frac{f[(1-\lambda)\xi + \lambda x]}{\lambda^{m+1}(\lambda - 1)} d\lambda]$$

for every $\psi \in \Psi$, $x \in B_r(\xi)$ and $m \in \mathbb{N}$. Since Ψ separates the points of F it follows that

$$f(x) - \sum_{k=0}^{m} P_k(x-\xi) = \frac{1}{2\pi i} \int_{|\lambda|=\rho} \frac{f[(1-\lambda)\xi + \lambda x]}{\lambda^{m+1}(\lambda - 1)} d\lambda$$

for every $x \in B_r(\xi)$ and $m \in \mathbb{N}$. Hence

$$\|f(x) - \sum_{k=0}^{m} P_k(x-\xi)\| \leq \frac{M}{\rho^m(\rho-1)}$$

for every $x \in B_r(\xi)$ and $m \in \mathbb{N}$. It follows that $\sum_{k=0}^{m} P_k(x-\xi)$ converges uniformly to f in $B_r(\xi)$. Q.E.D.

As a corollary to Proposition 6.1 we have

PROPOSITION 6.2 If F is a Banach space and $f: U \to F$ a mapping, the following are equivalent:

1) $f \in \mathcal{H}(U;F)$

2) $\psi \circ f \in \mathcal{H}(U;\mathbb{C})$ for every $\psi \in F'$, where F' is the topological dual of F.

PROOF: By the Banach-Steinhaus Theorem every weakly bounded subset of F is bounded, and so $\Psi = F'$ satisfies the conditions of Proposition 6.1. Thus Proposition 6.2 is a particular case of Proposition 6.1. Q.E.D.

PROPOSITION 6.3 Let E, F and G be normed spaces, F and G complete, U a non-empty open subset of E, and $f: U \to \mathcal{L}(F;G)$ a mapping. Then the following are equivalent:

a) $f \in \mathcal{H}(U;\mathcal{L}(F;G))$

b) the mapping $x \in U \to \omega[f(x)(y)] \in \mathbb{C}$ is holomorphic in U for every $y \in F$ and $\omega \in G'$.

PROOF: For $y \in F$ and $\omega \in G'$ the mapping

$$\psi_{\omega,y}: u \in \mathcal{L}(F;G) \longmapsto \omega[u(y)] \in \mathbb{C}$$

is linear and continuous. Let

$$\Psi = \{\psi_{\omega,y} : \omega \in G', y \in F\}.$$

If $x \in U$, $y \in F$ and $\omega \in G'$, then

$$(\psi_{\omega,y} \circ f)(x) = \psi_{\omega,y}[f(x)] = \omega[f(x)(y)],$$

and hence condition b) is equivalent to

c) $\psi_{\omega,y} \circ f \in \mathcal{H}(U;\mathbb{C})$ for every $\omega \in G'$, $y \in F$;

this can be written:

c') $\psi \circ f \in \mathcal{H}(U;\mathbb{C})$ for every $\psi \in \Psi$.

By hypothesis, G is complete, and so $\mathcal{L}(F;G)$ is also complete. Thus by Proposition 6.1 the equivalence of a) and c') is established once we have proved the following:

$B \subset \mathcal{L}(F;G)$ is bounded if and only if $\psi(B)$ is bounded for every $\psi \in \Psi$.

To say that $\psi(B)$ is bounded for every $\psi \in \Psi$ means that $\sup_{u \in B} |\omega[u(y)]| < \infty$ for every $\omega \in G'$ and $y \in F$; by the Banach-Steinhaus Theorem this is equivalent to $\sup_{u \in B} \|u(y)\| < \infty$ for every $y \in F$, and applying the Banach-Steinhaus Theorem once again (since F is complete), this in turn is equivalent to the condition that B is bounded in $\mathcal{L}(F;G)$. Q.E.D.

PROPOSITION 6.4 If F is complete and $f: U \to F'$ is a mapping, the following are equivalent:

a) $f \in \mathcal{H}(U;F')$

b) The mapping $x \in U \mapsto f(x)(y) \in \mathbb{C}$ is holomorphic in u for every $y \in F$.

PROOF: Take $G = \mathbb{C}$ in Proposition 6.3. Q.E.D.

CHAPTER 7

FINITE HOLOMORPHY AND GATEAUX HOLOMORPHY

DEFINITION 7.1 A mapping $f: U \to F$ is said to be finitely holomorphic if for every finite dimensional subspace S of E for which $S \cap U \neq \emptyset$ we have $f/S\cap U \in \mathcal{H}(U\cap S;F)$.

The set $\mathcal{H}_{fh}(U;F)$ of finitely holomorphic mappings of U into F is a complex vector space, the operations of addition and scalar multiplication being defined pointwise. We have:

a) $\mathcal{H}(U;F) \subset \mathcal{H}_{fh}(U;F)$ since the restrictions of polynomials from E into F to subspaces S of E are polynomials from S into F.

b) $f \in \mathcal{H}_{fh}(U;F)$ does not necessarily imply that $f \in \mathcal{H}(U;F)$. As we shall see presently, the condition $f \in \mathcal{H}(U;F)$ is equivalent to $f \in \mathcal{H}_{fh}(U;F)$ together with one other condition.

We shall denote by f_s the restriction of f to $S \cap U$.

REMARK 7.1 Let $f \in \mathcal{H}_{fh}(U;F)$, and let S be a finite dimensional subspace of E such that $S \cap U \neq \emptyset$. Then for $\xi \in U \cap S$ and $m \in \mathbb{N}$ we have $d^m f_s(\xi) \in \mathcal{L}_s(^m S;F)$ and $\hat{d}^m f_s(\xi) \in \mathcal{P}(^m S;F)$. If S_1 and S_2 are finite dimensional subspaces of E with $S_1 \subset S_2$ and $U \cap S_1 \neq \emptyset$, and

$\xi \in U \cap S_1 \subset U \cap S_2$, the following compatibility relations hold:

$$d^o f_{S_2}(\xi) = d^o f_{S_1}(\xi) = f(\xi); \quad d^m f_{S_2}(\xi)\Big|_{S_1} = d^m f_{S_1}(\xi) \quad \text{and}$$

$$\hat{d}^m f_{S_2}(\xi)\Big|_{S_1} = \hat{d}^m f_{S_1}(\xi) \quad \text{for every} \quad m \in \mathbb{N}, \quad m \geq 1.$$

This guarantees the existence for every $m \in \mathbb{N}$ of a mapping $\delta^m f(\xi) \in \mathcal{L}_{as}(^mE;F)$ such that $\delta^m f(\xi)\Big|_S = d^m f_S(\xi)$ for every finite dimensional subspace S of E. For $m = 0$, $\delta^o f(\xi) = f(\xi) \in \mathcal{L}_{as}(^oE;F)$. If $\hat{\delta}^m f(\xi)$ denotes the polynomial $\delta^m f(\xi)$ associated to $\delta^m f(\xi)$, then $\hat{\delta}^m f(\xi) \in \mathcal{P}_a(^mE;F)$ for every $\xi \in U$, $m \in \mathbb{N}$. In general we have no guarantee that $\delta^m f(\xi)$ or $\hat{\delta}^m f(\xi)$ will be continuous. However, if $f \in \mathcal{H}(U;F)$ then $d^m f(\xi) = \delta^m f(\xi)$ and $\hat{d}^m f(\xi) = \hat{\delta}^m f(\xi)$ for every $\xi \in U$ and $m \in \mathbb{N}$. This follows from the uniqueness of the Taylor expansion about the point ξ, and the fact that the restriction of a continuous symmetric m-linear mapping of E^m into F to a subspace S^m is itself a continuous symmetric m-linear mapping of S^m into F; thus we have $d^m f_S(\xi) = \delta^m f(\xi)\Big|_S$ for every finite dimensional subspace S of E, and hence $d^m f(\xi) = \delta^m f(\xi)$ for every $m \in \mathbb{N}$.

EXAMPLE 7.1 Every polynomial from E into F, continuous or not, is finitely holomorphic. This follows from the fact if A is any m-linear mapping of E^m into F, and S is a finite dimensional subspace of E, the restriction of A to S^m is m-linear and continuous.

If E is an infinite dimensional normed space, and E^*, E' are its algebraic and topological duals respectively, then $E^*\setminus E' \neq \emptyset$. For each $\varphi \in E^*\setminus E'$, $\varphi \in \mathcal{H}_{fh}(E;\mathbb{C})$ but $\varphi \notin \mathcal{H}(E;\mathbb{C})$.

PROPOSITION 7.1 Let E, F and G be complex normed spaces, U a non-empty open subset of F, $f \in \mathcal{H}(U;G)$, $\mu \in \mathcal{L}(E;F)$ and $a \in F$. Let μ_a be the continuous affine mapping of E into F defined by $\mu_a(x) = \mu(x) + a$, and suppose that $V = \mu_a^{-1}(U)$ is a non-empty open subset of E. Then $f\circ\mu_a \in \mathcal{H}(V;G)$, and for every $\zeta \in V$, $m \in \mathbb{N}$ we have

$$d^m(f\circ\mu_a)(\zeta) = d^m f[\mu_a(\zeta)]\circ\mu^m ,$$

$$\hat{d}^m(f\circ\mu_a)(\zeta) = \hat{d}^m f[\mu_a(\zeta)]\circ\mu$$

where $\mu^m\colon E^m \to F^m$ is given by

$$\mu^m(x_1,\ldots,x_m) = (\mu(x_1),\ldots,\mu(x_m)), \quad (x_1,\ldots,x_m) \in E^m.$$

<u>PROOF</u>: It is easy to see that if $A \in \mathcal{L}_s({}^mE;F)$ then $A\circ\mu^m \in \mathcal{L}_s({}^mE;G)$ and $\widehat{A\circ\mu^m} = \hat{A}\circ\mu \in \mathcal{P}({}^mE;G)$ for every m; thus the second equation follows from the first.

If $\xi = \mu_a(\zeta)$, $\zeta \in V$, we have $f(t) = \sum_{m=0}^{\infty} A_m(t-\xi)^m$ uniformly in a neighbourhood in U of $\xi = \mu_a(\zeta)$, where $A_m = \frac{1}{m!} d^m f(\xi)$, $m \in \mathbb{N}$. Thus

$$\begin{aligned} f[\mu_a(z)] &= \sum_{m=0}^{\infty} A_m[\mu_a(z)-\mu_a(\zeta)]^m \\ &= \sum_{m=0}^{\infty} A_m[\mu(z)-\mu(\zeta)]^m = \sum_{m=0}^{\infty} A_m[\mu(z-\zeta)]^m \\ &= \sum_{m=0}^{\infty} A_m\circ\mu^m(z-\zeta)^m , \end{aligned}$$

uniformly in a neighbourhood in V of ζ. Therefore

$f\circ\mu_a \in \mathcal{H}(V;G)$ and

$$\frac{1}{m!}\, d^m(f\circ\mu_a)(\zeta) = \frac{1}{m!}\, d^m f[\mu_a(\zeta)]\circ\mu^m$$

Q.E.D.

PROPOSITION 7.2 Let f be a mapping from U into F. The following conditions are equivalent:

1) f is holomorphic in U.

2) f is finitely holomorphic and continuous in U.

3) f is finitely holomorphic and locally bounded in U.

PROOF: The implications 1) $\Rightarrow$ 2) and 2) $\Rightarrow$ 3) are trivial. We shall prove 3) $\Rightarrow$ 1).

Let $f \in \mathcal{H}_{fh}(U;F)$ be locally bounded, and let $\xi \in U$. For each $x \in E$ we define a mapping $t_x\colon \mathbb{C} \to E$ by $t_x(\lambda) = \lambda x$, $\lambda \in \mathbb{C}$. Clearly $t_x \in \mathcal{L}(\mathbb{C};E)$. We denote by T_x the affine mapping $(t_x)_\xi$, defined, as in the preceeding Proposition by $(t_x)_\xi(\lambda) = \xi + \lambda x$, $\lambda \in \mathbb{C}$. The set $V_x = \{\lambda \in \mathbb{C} : \xi+\lambda x \in U\}$ is open, and $0 \in V_x$. If S_x is the subspace of E generated by ξ and x then $T_x(V_x) \subset U \cap S_x$; thus we may form the composition

$$V_x \subset \mathbb{C} \xrightarrow{T_x} U \cap S_x \xrightarrow{f} F.$$

Since S_x is finite dimensional and $f \in \mathcal{H}_{fh}(U;F)$, $f_{S_x} \in \mathcal{H}(U\cap S_x;F)$, and so by Proposition 7.1, $f_{S_x}\circ T_x \in \mathcal{H}(V_x;F)$. Using the fact that $0 \in V_x$, we have, again by Proposition 7.1,

$$(*) \qquad (f_{S_x} \circ T_x)(\lambda) = \sum_{m=0}^{\infty} \frac{1}{m!} d^m(f_{S_x} \circ T_x)(0)\lambda^m$$

$$= \sum_{m=0}^{\infty} \frac{1}{m!} [d^m f_{S_x}(T_x(0)) \circ t_x^m]\lambda^m$$

$$= \sum_{m=0}^{\infty} \frac{\lambda^m}{m!} d^m f_{S_x}(\xi)x^m$$

uniformly in some $\bar{D}_{r_x}(0)$. By the definition of the mappings $\delta^m f(\xi)$ and $\hat{\delta}^m f(\xi)$, this yields

$$f(\xi+\lambda x) = (f \circ T_x)(\lambda) = \sum_{m=0}^{\infty} \frac{\lambda^m}{m!} \delta^m f(\xi)x^m$$

$$= \sum_{m=0}^{\infty} \frac{\lambda^m}{m!} \hat{\delta}^m f(\xi)(x)$$

uniformly in some $\bar{D}_{r_x}(0)$.

Now since f is locally bounded in U there exist strictly positive real numbers ρ and M such that $\bar{B}_\rho(\xi) \subset U$ and $\sup_{\|t-\xi\| \leq \rho} \|f(t)\| \leq M$. If $x \in E$, $\|x\| \leq 1$, and $\lambda \in \mathbb{C}$, $|\lambda| \leq \rho$, then $\xi+\lambda x \in U$. Thus $\bar{D}_\rho(0) = \{\lambda \in \mathbb{C} : |\lambda| \leq \rho\} \subset V_x$ for every $x \in \bar{B}_1(0)$. Applying Proposition 4.2 to $f \circ T_x \in \mathcal{H}(V_x;F)$ we have

$$\frac{1}{m!} \hat{d}^m(f \circ T_x)(0)(1) = \frac{1}{2\pi i} \int_{|\lambda|=\rho} \frac{(f \circ T_x)(0+\lambda \cdot 1)}{\lambda^{m+1}} d\lambda$$

for every $x \in E$, $\|x\| \leq 1$ and $m \in \mathbb{N}$.

Thus

$$\frac{1}{m!} \hat{\delta}^m f(\xi)(x) = \frac{1}{2\pi i} \int_{|\lambda|=\rho} \frac{f(\xi+\lambda x)}{\lambda^{m+1}} d\lambda$$

for every $x \in E$, $\|x\| \leq 1$ and $m \in \mathbb{N}$. This shows that

$$\left\|\frac{1}{m!}\hat{\delta}^m f(\xi)\right\| \leq \frac{M}{\rho^m} < \infty$$

for every $m \in \mathbb{N}$. Thus $\hat{\delta}^m f(\xi)$ is continuous, that is, $\hat{\delta}^m f(\xi) \in \mathcal{P}(^m E;F)$, for every $m \in \mathbb{N}$.

Let σ and r be real numbers such that $\sigma > 1$, $r > 0$ and $\sigma r \leq \rho$. We shall prove that the series $\sum_{m=0}^{\infty} \frac{1}{m!}\hat{\delta}^m f(\xi)(y-\xi)$ converges uniformly to $f(y)$ in $\bar{B}_r(\xi)$.

(Note that $\bar{B}_r(\xi) \subset \bar{B}_\rho(\xi) \subset U$). Let $y \in \bar{B}_r(\xi)$. Taking $x = y-\xi$, we know that $f \circ T_x \in \mathcal{H}(V_x;F)$. Applying Proposition 4.5 to the points 0 and 1 of V_x, since $\lambda \in V_x$ for $\lambda \in \mathbb{C}$, $|\lambda| \leq \sigma$, we have

$$(f\circ T_{y-\xi})(1) - \sum_{k=0}^{m} \frac{1}{k!}\hat{d}^k(f\circ T_{y-\xi})(0)(1)$$

$$= \frac{1}{2\pi i}\int_{|\lambda|=\sigma} \frac{f[(1-\lambda)\xi+\lambda y]}{\lambda^{m+1}(\lambda-1)}\, d\lambda,$$

for every $y \in \bar{B}_r(\xi)$, $m \in \mathbb{N}$, Thus

$$f(y) - \sum_{k=0}^{m} \frac{1}{k!}\hat{\delta}^k f(\xi)(y-\xi) = \frac{1}{2\pi i}\int_{|\lambda|=\sigma} \frac{f[(1-\lambda)\xi+\lambda y]}{\lambda^{m+1}(\lambda-1)}\, d\lambda$$

for every $y \in \bar{B}_r(\xi)$, $m \in \mathbb{N}$. It follows that

$$\left\|f(y) - \sum_{k=0}^{m} \frac{1}{k!}\hat{\delta}^k f(\xi)(y-\xi)\right\| \leq \frac{M}{\sigma^m(\sigma-1)}$$

for every $y \in \bar{B}_r(\xi)$, $m \in \mathbb{N}$. Since $\sigma > 1$ we conclude that the series $\sum_{m=0}^{m} \frac{1}{m!}\hat{\delta}^m f(\xi)(y-\xi)$ converges uniformly to $f(y)$ in $\bar{B}_r(\xi)$. Q.E.D.

DEFINITION 7.2 A mapping $f: U \to F$ is said to be G-holomorphic (Gateaux-holomorphic) in U if for every $a \in U$ and $b \in E$ the mapping

$$\lambda \in \{\lambda \in \mathbb{C} : a+\lambda b \in U\} \subset \mathbb{C} \longmapsto f(a+\lambda b) \in F$$

is holomorphic in the open subset $\{\lambda \in \mathbb{C} : a+\lambda b \in U\}$ of $\mathbb{C}$.

REMARK 7.2 This definition is equivalent to the following statement:

a) If x and y are points in E such that the set $A = \{\lambda \in \mathbb{C} : x+\lambda y \in U\}$ is not empty, then the mapping

$$\lambda \in A \subset \mathbb{C} \mapsto f(x+\lambda y) \in F$$

is holomorphic in the open set A.

It is clear that a) implies the condition given in Definition 7.2. Conversely, suppose that x and y are points in E such that the mapping

$$\lambda \in \{\lambda \in \mathbb{C} : x+\lambda y \in U\} \longmapsto f(x+\lambda y) \in F$$

is not holomorphic at a point λ_o of its domain. Then $a = x+\lambda_o y \in U$, and the mapping

$$\lambda \in \{\lambda \in \mathbb{C} : a+\lambda y \in U\} \subset \mathbb{C} \mapsto f(a+\lambda y)$$

is not holomorphic at the point $\lambda = 0$. For, suppose there existed $b_m \in F$, $m \in \mathbb{N}$ and $\rho > 0$ such that

$$(*) \qquad f(a+\lambda y) = f(a) + \sum_{m=1}^{\infty} b_m \lambda^m$$

uniformly in $\bar{B}_\rho(0) = \{\lambda \in \mathbb{C} : |\lambda| \leq \rho\}$. We may write (*) as:

$$f(x+\lambda_o y+\lambda y) = f(x+(\lambda_o+\lambda)y) = f(x+\lambda_o y) + \sum_{m=1}^{\infty} b_m\lambda^m,$$

whence $f(x+\lambda_o y) + \sum_{m=1}^{\infty} b_m\lambda^m$ converges to $f(x+(\lambda_o+\lambda)y)$ uniformly in $B_\rho(0)$. Thus the function

$$\lambda \in \{\lambda \in \mathbb{C} : x+\lambda y \in U\} \subset \mathbb{C} \mapsto f(x+\lambda y) \in F$$

is holomorphic at λ_o, contradicting our hypothesis. It follows that Definition 7.2 implies condition a).

Let U be a non-empty open subset of $\mathbb{C}^p$, $p \in \mathbb{N}$, $p \geq 1$, F a complex normed space, and $\{e_1,\ldots,e_p\}$ the canonical basis for $\mathbb{C}^p$. We denote by J_k the mapping

$$\lambda \in \mathbb{C} \mapsto \lambda e_k, \qquad 1 \leq k \leq p,$$

$z^k = (z_1,\ldots,z_{k-1},0,z_{k+1},\ldots,z_p) \in \mathbb{C}^p$. Then for $\lambda \in \mathbb{C}$ and $k = 1,2,\ldots,p$ we have

$$(z_1,\ldots,z_{k-1},\lambda,z_{k+1},\ldots,z_p) = z^k + \lambda e_k = z^k + J_k(\lambda).$$

$J_{z,k}$ denotes the continuous affine linear mapping

$$\lambda \in \mathbb{C} \mapsto J_{z,k}(\lambda) = z^k + J_k(\lambda) = z^k + \lambda e_k \in \mathbb{C}^p.$$

DEFINITION 7.3 Let U be a non-empty open subset of $\mathbb{C}^p$. A mapping $f: U \to F$ is separately holomorphic in U if it satisfies the following condition: For every $z = (z_1,\ldots,z_p) \in \mathbb{C}^p$, and $k = 1,2,\ldots,p$ such that $U_{z,k} = J_{z,k}^{-1}(U) \neq \emptyset$ we have $f \circ J_{z,k} \in \mathcal{H}(U_{z,k};F)$.

We remark that this condition is equivalent to the requirement that for every $z = (z_1, \ldots, z_p) \in \mathbb{C}^p$ and $k = 1,2,\ldots,p$, the restriction of f to the section of U determined by $z_1, \ldots, z_{k-1}, z_{k+1}, \ldots, z_p$ is holomorphic.

PROPOSITION 7.3 (Theorem of Hartogs): Let U be a non-empty open subset of $\mathbb{C}^p$, F a complex normed space, and f a mapping of U into F. The following are equivalent:

a) f is holomorphic in U.

b) f is separately holomorphic in U.

REMARK 7.3 The proof of the theorem of Hartogs may be found in [60].

PROPOSITION 7.4 Let U be a non-empty open subset of E and f a mapping of U into F. The following are equivalent:

1) f is G-holomorphic in U.

2) f is finitely holomorphic in U.

PROOF: 1) $\Rightarrow$ 2). Let S be a finite dimensional subspace of E such that $V = S \cap U \neq \emptyset$ and let $\{a_1, \ldots, a_p\}$ be a basis for S. Since the mapping

$$\varphi: (z_1, \ldots, z_p) \in \mathbb{C}^p \mapsto \sum_{k=1}^{p} z_k a_k \in S$$

is a homeomorphism and an isomorphism between the vector spaces $\mathbb{C}^p$ and S the set $W = \varphi^{-1}(V)$ is a non-empty open subset of $\mathbb{C}^p$. Consider the mapping

$$g: (z_1, \ldots, z_p) \in W \mapsto f(\sum_{k=1}^{p} z_k a_k) \in F.$$

We have $f_S = g \circ \varphi^{-1}$ and so, by Proposition 7.1, to show that

$f_S \in \mathcal{H}(U \cap S;F)$ it suffices to prove that $g \in \mathcal{H}(W;F)$. By Proposition 7.3 this is equivalent to g being separately holomorphic.

For $z = (z_1,\dots,z_p) \in \mathbb{C}^p$ and $k \in \mathbb{N}$, $1 \leq k \leq p$, such that

$$\begin{aligned} W_{z,k} &= \{\lambda \in \mathbb{C} : (z_1,\dots,z_{k-1},\lambda,z_{k+1},\dots,z_p) \in W\} \\ &= \{\lambda \in \mathbb{C} : z^k+\lambda e_k \in W\} \neq \emptyset, \end{aligned}$$

we must show that the mapping

$$\lambda \in W_{z,k} \mapsto g(z^k+\lambda e_k) \in F$$

is holomorphic in $W_{z,k}$.

We have $g(z^k+\lambda e_k) = (f\circ\varphi)(z^k+\lambda e_k) = f[\varphi(z^k)+\lambda a_k]$, and from the definition of a G-holomorphic mapping and Remark 7.2, we know that the mapping

$$\lambda \in W_{z,k} \mapsto f[\varphi(z^k)+\lambda a_k] \in F$$

is holomorphic in $W_{z,k}$.

2) $\Rightarrow$ 1). Let $a \in U$, $b \in E$. If S is the subspace of E generated by a and b then $a \in U \cap S$ and by 2), $f_S \in \mathcal{H}(U \cap S;F)$. Applying Proposition 7.1 to the composition

$$\lambda \in \{\lambda \in \mathbb{C} : a+\lambda b \in U\} \mapsto a+\lambda b \in U \cap S \mapsto f(a+\lambda b) \in F,$$

this mapping is holomorphic. Thus f is G-holomorphic.

Q.E.D.

COROLLARY 7.1 Let f be a mapping of U into F. The following are equivalent:

a) f is holomorphic in U.

b) f is G-holomorphic and continuous in U.

c) f is G-holomorphic and locally bounded in U.

PROOF: This corollary is an immediate consequence of Propositions 7.2 and 7.4. Q.E.D.

CHAPTER 8

TOPOLOGIES ON SPACES OF HOLOMORPHIC MAPPINGS

DEFINITION 8.1 A non-empty subset X of E is said to be U-bounded if $X \subset U$, X is bounded and the distance from X to the boundary of U is positive.

DEFINITION 8.2 $f \in \mathcal{H}(U;F)$ is said to be of bounded type if f is bounded on every U-bounded subset of U.

$\mathcal{H}_b(U;F)$ will denote the vector space of holomorphic mappings of bounded type from U into F.

If X is a U-bounded subset of E, the mapping $p_X: f \in \mathcal{H}_b(U;F) \mapsto p_X(f) = \sup\{\|f(x)\| : x \in X\} \in \mathbb{R}$ is a seminorm on $\mathcal{H}_b(U;F)$. The family of seminorms p_X where X ranges over all the U-bounded subsets of E, defines a separated locally convex topology on $\mathcal{H}_b(U;F)$, which we denote by τ_b. This topology is known as the natural topology on $\mathcal{H}_b(U;F)$.

PROPOSITION 8.1 If F is complete then $\mathcal{H}_b(U;F)$, with the topology τ_b, is a Fréchet space.

PROOF: For $n = 1,2,...,$ let

$$X_n = \{x \in U : \|x\| \leq n \text{ and } \operatorname{dist}(x;\partial U) \geq \frac{1}{n}\}.$$

There exists $n_o \in \mathbb{N}$ such that for $n \geq n_o$, X_n is non-empty and hence U-bounded. For

$$U = \bigcup_{n=1}^{\infty} U_n \qquad \text{and} \qquad U = \bigcup_{m=1}^{\infty} V_m$$

where $U_n = \{x \in U : \|x\| \leq n\}$ and $V_m = \{x \in U : \text{dist}(x;\partial U) \geq \geq \frac{1}{m}\}$ there exist positive integers m_o and n_o such that $U_{n_o} \cap V_{m_o} \neq \emptyset$. It follows that there is an index $n_o \geq 1$ for which $X_{n_o} \neq \emptyset$.

Moreover, every U-bounded subset of E is contained in X_n for n sufficiently large. Thus the sequence $\{p_{X_n}\}$ $n = 1,2,\ldots$ determines the topology τ_b. Therefore τ_b is a metrizable topology.

Let $\{f_m\}_{m \in N}$ be a τ_b-Cauchy sequence in $\mathcal{H}_b(U;F)$. Since every compact subset of U is U-bounded, $\{f_m\}_{m \in N}$ is a τ-Cauchy sequence, where τ is the topology induced on $\mathcal{H}_b(U;F)$ by the compact-open topology of $C(U;F)$. By Proposition 4.6, $\mathcal{H}(U;F)$ is closed in $C(U;F)$ for the compact-open topology, and hence $\{f_m\}_{m \in \mathbb{N}}$ converges to some $f \in \mathcal{H}(U;F)$ for this topology. It follows that $f(x) = \lim_{m \to \infty} f_m(x)$ for every $x \in U$. Now if $X \subset U$ is a U-bounded set, and $\varepsilon > 0$, then since $\{f_m\}_{m \in \mathbb{N}}$ is a τ_b-Cauchy sequence, there exists $m_o \in \mathbb{N}^* = \{1,2,\ldots\}$ such that

$$\|f_m(x) - f_n(x)\| < \varepsilon \quad \text{for every} \quad x \in X, \quad m,n \geq m_o.$$

Therefore

$$(*) \qquad \|f(x) - f_n(x)\| \leq \varepsilon \quad \text{for every} \quad x \in X, \quad n \geq m_o,$$

which implies that

$$\sup_{x\in X} \|f(x)\| \le \sup_{x\in X} \|f_{m_o}(x)\| + \varepsilon < +\infty .$$

Thus $f \in \mathcal{H}_b(U;F)$; and (*) shows that $\{f_m\}_{m\in\mathbb{N}}$ converges to f for the topology τ_b. Q.E.D.

DEFINITION 8.3 $\mathcal{H}_B(U;F)$ denotes the vector space of holomorphic mappings of U into F which are bounded on U. The mapping

$$f \in \mathcal{H}_B(U;F) \mapsto \|f\| = \sup_{x\in U} \|f(x)\| \in \mathbb{R}$$

is a norm on $\mathcal{H}_B(U;F)$.

PROPOSITION 8.2 If F is complete then $\mathcal{H}_B(U;F)$ with the norm above is complete.

PROOF: Let $\{f_m\}_{m\in\mathbb{N}}$ be a Cauchy sequence in $(\mathcal{H}_B(U;F), \| \|)$. Then, given $\varepsilon > 0$, there exists $n_o \in \mathbb{N}$ such that for $m,n \ge n_o$,

$$(*) \qquad \|f_m - f_n\| = \sup_{x\in U} \|f_m(x) - f_n(x)\| < \varepsilon .$$

As in the proof of Proposition 8.1, it follows that $\{f_m\}_{m\in\mathbb{N}}$ converges to some $f \in \mathcal{H}(U;F)$ for the topology induced by the compact-open topology on $C(U;F)$. We must show that $f \in \mathcal{H}_B(U;F)$. Since $\lim_{m\to\infty} f_m(x) = f(x)$ for every $x \in U$, it follows from (*) that there exists $n_o \in \mathbb{N}$ such that

$$(**) \qquad \sup_{x\in U} \|f(x) - f_n(x)\| \le 1$$

for every $n \ge n_o$. Hence

$$\sup_{x\in U} \|f(x)\| \le \sup_{x\in U} \|f_{n_o}(x)\| + 1 < +\infty .$$

Therefore $f \in \mathcal{H}_B(U;F)$; and we conclude from (*) that $f_m \to f$ in $(\mathcal{H}_B(U;F), \| \|)$. Q.E.D.

We now consider the various ways of topologizing $\mathcal{H}(U;F)$.

DEFINITION 8.4 τ_o will denote the topology on $\mathcal{H}(U;F)$ induced by the compact-open topology of $C(U;F)$. τ_o is then a separated locally convex topology on $\mathcal{H}(U;F)$, defined by the seminorms

$$p_K: f \in \mathcal{H}(U;F) \mapsto p_K(f) = \sup_{x \in K} \|f(x)\|$$

as K ranges over the compact subsets of U.

DEFINITION 8.5 If $m \in \mathbb{N}$, τ_m will denote the separated locally convex topology on $\mathcal{H}(U;F)$ defined by the seminorms

$$p_{K,n}: f \in \mathcal{H}(U;F) \mapsto p_{K,n}(f) = \sup_{x \in K} \|\frac{1}{m!} \hat{d}^n f(x)\|$$

as K ranges over the compact subsets of U, and n over the set $\{0,1,2,\ldots,m\}$.

DEFINITION 8.6 τ_∞ denotes the separated locally convex topology on $\mathcal{H}(U;F)$ defined by the seminorms

$$p_{K,n}: f \in \mathcal{H}(U;F) \mapsto p_{K,n}(f) = \sup_{x \in K} \|\frac{1}{n!} \hat{d}^n f(x)\|$$

as K ranges over the compact subsets of U, and n over $\mathbb{N}$.

REMARK 8.1: a) It follows from Proposition 1.3 that the topologies defined in 8.5 and 8.6 are unchanged if the symbol $\hat{d}$ is replaced by d.

b) It is obvious that $\tau_o \leq \tau_m \leq \tau_{m+1} \leq \tau_\infty$ for every $m \in \mathbb{N}$.

We introduce next the topology τ_ω on $\mathcal{H}(U;F)$.

DEFINITION 8.7 A seminorm p on $\mathcal{H}(U;F)$ is said to be ported by a compact subset K of U if for every neighbourhood V of K contained in U there is a real number $C(V) > 0$ such that

$$p(f) \leq C(V) \sup_{x \in V} \|f(x)\| \quad \text{for every} \quad f \in \mathcal{H}(U;F).$$

τ_ω denotes the separated locally convex topology on $\mathcal{H}(U;F)$ defined by the seminorms p which are ported by some compact subset K of U (K may vary with p).

REMARK 8.2: The topology τ_ω was introduced by Nachbin in [91], having as a motivation Martineau's concept of a linear analytic functional ported by a compact subset. See Martineau [76].

REMARK 8.3: If the seminorm p on $\mathcal{H}(U;F)$ is ported by a compact $K_1 \subset U$ and if K_2 is another compact set with $K_1 \subset K_2 \subset U$, then p is also ported by K_2.

On the other hand, if the seminorm p is ported by each of two compact subsets of U, K_1 and K_2, it is not, in general, true that p is ported by $K_1 \cap K_2$.
In other words, a seminorm which is ported by a compact set does not, in general, possess a minimal compact porting set.

EXAMPLE 8.1 Let $U = E = F = \mathbb{C}$, and for $\xi \in \mathbb{C}$, let $p(f) = |f(\xi)|$, $f \in \mathcal{H}(\mathbb{C},\mathbb{C})$. It is clear that p is a semi norm on $\mathcal{H}(\mathbb{C},\mathbb{C})$. By the maximum modulus theorem, if $a \in C$ and $|a-\xi| < r$, then $p(f) = |f(\xi)| \le \sup\{|f(x)| : x \in S_r(a)\}$ for every $f \in \mathcal{H}(\mathbb{C},\mathbb{C})$, where $S_r(a) = \{x \in \mathbb{C} : |x-a| = r\}$. Therefore p is ported by the compact set $K_1 = S_r(a)$. If r' is a real number such that $|\xi-a| < r' < r$, and $K_2 = S_{r'}(a)$, then p is also ported by K_2. However, p cannot be ported by $K_1 \cap K_2$ since $K_1 \cap K_2 = \emptyset$, and p is not the zero seminorm.

We note that the unique seminorm ported by the empty set is the seminorm which is identically zero.

REMARK 8.4: Let $\mathfrak{F}$ be a filter in $\mathcal{H}(U;F)$ and $X \subset U$. We say that $\mathfrak{F}$ converges to $f \in \mathcal{H}(U;F)$ uniformly on X if, for every real number $\varepsilon > 0$ there exists $M = M(\varepsilon) \in F$ such that if $g \in M$ then $\|f(x)-g(x)\| \le \varepsilon$ for every $x \in X$.

PROPOSITION 8.3 Let p be a seminorm on $\mathcal{H}(U;F)$ and let K be a compact subset of U. The following are equivalent:

a) p is ported by K.

b) If $\mathfrak{F}$ is a filter in $\mathcal{H}(U;F)$ for which there exists a neighbourhood V of K contained in U such that $\mathfrak{F}$ converges to zero uniformly on V, the filter base $p(\mathfrak{F})$ in $\mathbb{R}_+ = \{x \in \mathbb{R} : x \ge 0\}$ converges to 0.

PROOF: a) $\Rightarrow$ b). Let $\mathfrak{F}$ be a filter in $\mathcal{H}(U;F)$ which converges uniformly to zero on a neighbourhood V of K contained in U. The family $\mathcal{U}'$ of sets of the form $W_\varepsilon = [0,\varepsilon]$, ε ranging over the set of positive real numbers, is a base for the filter $\mathcal{V}$ of neighbourhoods of 0 in $\mathbb{R}_+$. Let $C(V) > 0$ be the number given by Definition 8.7, and let $\varepsilon > 0$. Since $\mathfrak{F}$ converges to zero uniformly on V there exists $M \in \mathfrak{F}$ such that if $g \in M$, then $\|g(x)\| < \varepsilon/C(V)$ for every $x \in V$. Therefore, using Definition 8.7, $p(g) \leq C(v)\cdot\varepsilon/C(V) = \varepsilon$ for $g \in M$. Hence $p(M) = \{p(g) : g \in M\} \subset W_\varepsilon$, which shows that the filter generated by $p(\mathfrak{F})$ is finer than $\mathcal{V}$. Therefore $p(\mathfrak{F})$ converges to 0.

b) $\Rightarrow$ a). Suppose that p satisfies b) but is not ported by K. We shall construct a filter $\mathfrak{F}$ in $\mathcal{H}(U;F)$ and a neighbourhood V of K contained in U such that $\mathfrak{F}$ converges to zero uniformly in V while $p(\mathfrak{F})$ does not converge to zero, contradicting b).

If p is not ported by K, there is a neighbourhood V of K contained in U such that for every $k \in \mathbb{N}$, $k \geq 1$, there is an $f_k \in \mathcal{H}(U;F)$ for which

$$(*) \qquad p(f_k) > k \sup_{x\in V} \|f_k(x)\|.$$

Then $g_k = f_k/p(f_k) \in \mathcal{H}(U;F)$ and $p(g_k) = 1$ for every $k \in \mathbb{N}$, $k \geq 1$. Let $\mathfrak{F}$ be the filter in $\mathcal{H}(U;F)$ generated by the image under the mapping $k \in \mathbb{N}-\{0\} \mapsto g_k \in \mathcal{H}(U;F)$ of the Fréchet filter in $\mathbb{N}-\{0\}$, which has as base the sets $N_k = \{n \in \mathbb{N} : n \geq k\}$, $k \in \mathbb{N}$, $k \geq 1$. From (*) we have

$$(**) \qquad \|g_k(x)\| = \frac{\|f_k(x)\|}{p(f_k)} < \frac{1}{k}$$

for every $k \in \mathbb{N}-\{0\}$ and every $x \in V$. (**) shows that $\mathfrak{F}$ converges to zero uniformly on V. However, $p(\mathfrak{F})$ does not converge to 0 in $\mathbb{R}_+$, since $p(g_k) = 1$ for every $k \in \mathbb{N}$, $k \geq 1$. This contradicts b), and so p is ported by K.

Q.E.D.

EXAMPLE 8.2 Let K be a compact subset of U, and let p be defined by

$$p: f \in \mathcal{H}(U;F) \mapsto p(f) = \sup_{x \in K} \|f(x)\|.$$

It is easy to see that the seminorm p is ported by K.

EXAMPLE 8.3 Let K be a compact subset of U and let $\alpha = \{\alpha_m\}_{m \in \mathbb{N}}$ be a sequence of non-negative real numbers such that $\lim_{m \to \infty} (\alpha_m)^{1/m} = 0$. Then the mapping

$$p_{K,\alpha}: f \in \mathcal{H}(U;F) \mapsto p_{K,\alpha}(f) = \sum_{m=0}^{\infty} \alpha_m \sup_{x \in K} \|\frac{1}{m!} \hat{d}^m f(x)\| \in \mathbb{R}$$

is a seminorm on $\mathcal{H}(U;F)$ which is ported by K. To see this, let V be a neighbourhood of K contained in U. Then there is a real number r, $r > 0$, such that $V_r = \bigcup_{x \in K} \bar{B}_r(x) \subset V$. Applying the Cauchy inequalities, Proposition 4.4, we have

$$\sup_{x \in K} \|\frac{1}{m!} \hat{d}^m f(x)\| \leq \frac{1}{r^m} \sup_{t \in V} \|f(t)\|,$$

and so

$$c(V) = \sum_{m=0}^{\infty} \frac{\alpha_m}{r^m} = \sum_{m=0}^{\infty} \left(\frac{\alpha_m^{1/m}}{r}\right)^m < +\infty.$$

Then

$$p_{K,\alpha}(f) = \sum_{m=0}^{\infty} \alpha_m \sup_{x \in V} \|\frac{1}{m!} \hat{d}^m f(x)\| \leq c(V) \sup_{t \in V} \|f(t)\|.$$

It follows that if $p_{K,\alpha}$ is a seminorm on $\mathcal{H}(U;F)$, then it is ported by K. We show that $p_{K,\alpha}(f) < +\infty$ for every $f \in \mathcal{H}(U;F)$, the remaining properties of a seminorm being immediate. Since $f \in \mathcal{H}(U;F)$ is locally bounded, there is a neighbourhood V_f of K in U such that $\sup\{\|f(t)\| : t \in V_f\} < +\infty$. Therefore

$$p_{K,\alpha}(f) \leq C(V_f) \sup_{t \in V_f} \|f(t)\| < +\infty.$$

REMARK 8.5: We may replace $\hat{d}$ by d in this example without altering the conclusion, since, if $\{\alpha_m\}_{m\in\mathbb{N}}$ is a sequence of non-negative real numbers for which $\lim_{m\to\infty} \alpha_m^{1/m} = 0$, and $\beta_m = \alpha_m(m^m/m!)$, $m \in \mathbb{N}$, then $\beta_m \geq 0$ for every $m \in \mathbb{N}$, and

$$\lim_{m\to\infty} (\beta_m)^{1/m} = (\lim_{m\to\infty} (\alpha_m)^{1/m})e = 0.$$

PROPOSITION 8.4 $\tau_\infty \leq \tau_\omega$ on $\mathcal{H}(U;F)$.

PROOF: Let $K \subset U$ be compact and $n \in \mathbb{N}$. Consider the seminorm $p_{K,n}$ on $\mathcal{H}(U;F)$, defined by

$$p_{K,n}(f) = \sup_{x\in K} \|\hat{d}^n f(x)\|, \qquad f \in \mathcal{H}(U;F).$$

If we define the sequence $\{\alpha_m\}_{m\in\mathbb{N}}$ by $\alpha_m = 0$ if $m \neq n$ and $\alpha_n = 1$, then Example 8.3 shows that $p_{K,n}$ is ported by K. Since τ_∞ is defined by the family of seminorms $p_{K,n}$, it follows that $\tau_\infty \leq \tau_\omega$. Q.E.D.

REMARK 8.6: Example 8.3 gives us a method for constructing an infinite number of seminorms on $\mathcal{H}(U;F)$ which are ported by compact sets. Denoting by Γ the set of seminorms which arise in this way, a natural question to ask is whether Γ is a fundamental system of seminorms for the topology τ_ω. In other words, if p is a τ_ω-continuous seminorm on $\mathcal{H}(U;F)$, does there exist $p_{K,\alpha} \in \Gamma$ such that $p \leq p_{K,\alpha}$? A positive answer to this question would give us an explicit, manageable expression for the seminorms on $\mathcal{H}(U;F)$ which are ported by compact subsets of U. A negative answer would imply the existence of another locally convex topology τ_Γ on $\mathcal{H}(U;F)$, defined by the family Γ of seminorms. In this case we would have $\tau_\infty \leq \tau_\Gamma \leq \tau_\omega$.

PROPOSITION 8.5 Let p be a seminorm on $\mathcal{H}(U;F)$. The following are equivalent:

a) p is ported by a compact subset K of U.

b) For every real number ε, $\varepsilon > 0$, there exists a real number $c(\varepsilon)$, $c(\varepsilon) > 0$, such that

$$p(f) \leq c(\varepsilon) \sum_{m=0}^{\infty} \varepsilon^m \sup_{x \in K} \left\| \frac{1}{m!} \hat{d}^m f(x) \right\|$$

for every $f \in \mathcal{H}(U;F)$.

c) For every real number $\varepsilon > 0$, and for every open neighbourhood V of K contained in U there exists a real number $c(\varepsilon, V) > 0$ such that

$$p(f) \leq c(\varepsilon, V) \sum_{m=0}^{\infty} \varepsilon^m \sup_{x \in V} \left\| \frac{1}{m!} \hat{d}^m f(x) \right\|$$

for every $f \in \mathcal{H}(U;F)$.

The proof of this proposition requires the following two lemmas.

LEMMA 8.1 Let $f \in \mathcal{H}(U;F)$ and $\xi \in U$. If $\rho \in \mathbb{R}$, $\rho > 0$ is such that $\bar{B}_{2\rho}(\xi) \subset U$ then

$$\left\|\frac{1}{k!}\hat{d}^k f(x)\right\| \leq \sum_{m=k}^{\infty} 2^m \rho^{m-k} \left\|\frac{1}{m!}\hat{d}^m f(\xi)\right\|$$

for every $x \in \bar{B}_\rho(\xi)$, $k = 0,1,2,\ldots$

PROOF OF LEMMA 8.1: By Proposition 5.4, $\frac{1}{k!}\hat{d}^k f \in \mathcal{H}(U;\mathcal{P}(^kE;F))$ for every $k \in \mathbb{N}$. Since $\bar{B}_{2\rho}(\xi)$ is ξ-balanced, the Corollary 5.1 shows that

1) $$\frac{1}{k!}\hat{d}^k f(x) = \sum_{m=k}^{\infty} \frac{1}{k!}\hat{d}^k P_m(x-\xi)$$

for $x \in B_{2\rho}(\xi)$, $k = 0,1,2,\ldots$, the convergence being pointwise with respect to the norm of $\mathcal{P}(^kE;F)$, where $P_m = \frac{1}{m!}\hat{d}^m f(\xi)$.

For $m \in \mathbb{N}$, $m \geq k$, consider the mapping $P_m' : U \to F$ defined by $P_m'(y) = P_m(y-\xi)$. $P_m' \in \mathcal{H}(U;F)$ and so, applying the Cauchy inequalities,

$$\left\|\frac{1}{k!}\hat{d}^k P_m(x-\xi)\right\| = \left\|\frac{1}{k!}\hat{d}^k P_m'(x)\right\|$$

$$\leq \frac{1}{\rho^k} \sup_{\|y-x\|=\rho} \|P_m'(y)\| = \frac{1}{\rho^k} \sup_{\|y-x\|=\rho} \|P_m(y-\xi)\|$$

for $x \in \bar{B}_\rho(\xi)$, $k = 0,1,\ldots$ and $m \geq k$. But if $\|x-y\| = \rho$ and $x \in \bar{B}_\rho(\xi)$ then

$$\|P_m(y-\xi)\| \leq \|P_m\| \, \|y-\xi\|^m \leq \|P_m\| \, (\|y-x\| + \|x-\xi\|)^m$$

$$\leq \|P_m\| \, (2\rho)^m.$$

Therefore

2) $$\left\|\frac{1}{k!}\hat{d}^k P_m(x-\xi)\right\| \leq 2^m \|P_m\| \rho^{m-k}$$

for $k = 0,1,\ldots$, $m \in \mathbb{N}$, $m \geq k$ and $x \in \bar{B}_\rho(\xi)$.

From 1) and 2) we conclude that

$$\left\|\frac{1}{k!}\hat{d}^k f(x)\right\| \leq \sum_{m=k}^{\infty} 2^m \rho^{m-k} \left\|\frac{1}{m!}\hat{d}^m f(\xi)\right\|$$

for $k = 0,1,\ldots$ and $x \in \bar{B}_\rho(\xi)$. Q.E.D.

LEMMA 8.2 Let $f \in \mathcal{H}(U;F)$ and $X \subset U$. If $\rho \in \mathbb{R}$, $\rho > 0$ is such that $\bar{B}_{2\rho}(X) \subset U$, then for every real number $\varepsilon > 0$, we have

$$\sum_{k=0}^{\infty} \varepsilon^k \sup_{x\in\bar{B}_\rho(X)} \left\|\frac{1}{k!}\hat{d}^k f(x)\right\| \leq$$

$$\leq \sum_{m=0}^{\infty} [2(\rho+\varepsilon)]^m \sup_{x\in X} \left\|\frac{1}{m!}\hat{d}^m f(x)\right\| .$$

PROOF OF LEMMA 8.2: Let $y \in \bar{B}_\rho(X)$; then there exists $x \in X$ such that $y \in \bar{B}_\rho(x)$, and so by Lemma 8.1,

$$\left\|\frac{1}{k!}\hat{d}^k f(y)\right\| \leq \sum_{m=k}^{\infty} 2^m \rho^{m-k} \left\|\frac{1}{m!}\hat{d}^m f(x)\right\| \leq$$

$$\leq \sum_{m=k}^{\infty} 2^m \rho^{m-k} \sup_{x\in X} \left\|\frac{1}{m!}\hat{d}^m f(x)\right\|$$

for $k = 0,1,\ldots$. Therefore

$$\sup_{y\in\bar{B}_\rho(X)} \left\|\frac{1}{k!}\hat{d}^k f(y)\right\| \leq$$

$$\leq \sum_{m=k}^{\infty} 2^m \rho^{m-k} \sup_{x\in X} \left\|\frac{1}{m!}\hat{d}^m f(x)\right\|$$

for $k = 0,1,\ldots$.

Let

$$M_k = \sup_{y\in B_\rho(X)} \left\|\frac{1}{k!}\hat{d}^k f(y)\right\|, \qquad k \in \mathbb{N}$$

and

$$M'_m = \sup_{x\in X} \left\|\frac{1}{m!}\hat{d}^m f(x)\right\|, \qquad m \in \mathbb{N}.$$

Then $M_k \leq \sum_{m\geq k} 2^m \rho^{m-k} M'_m$, and hence for every $\varepsilon \in R$, $\varepsilon > 0$ and $k \in \mathbb{N}$ we have

$$\sum_{k=0}^{\infty} \varepsilon^k M_k \leq \sum_{k=0}^{\infty} \left(\sum_{m=k}^{\infty} \varepsilon^k 2^m \rho^{m-k} M'_m \right)$$

which is the statement of the Lemma 8.2. Q.E.D.

PROOF OF PROPOSITION 8.5:

a) $\Rightarrow$ c). Suppose that p is ported by K. Let $\varepsilon \in \mathbb{R}$, $\varepsilon > 0$, and let V be an open neighbourhood of K contained in U. There is a real number $c(V) > 0$ such that

1) $$p(f) \leq c(V) \sup_{x\in V} \|f(x)\|$$

for every $f \in \mathcal{H}(U;F)$. But since

$$\sup_{x\in V} \|f(x)\| \leq \sum_{m=0}^{\infty} \varepsilon^m \sup_{x\in V} \left\|\frac{1}{m!} \hat{d}^m f(x)\right\|$$

for $f \in \mathcal{H}(U;F)$, c) follows from 1) with $c(\varepsilon,V) = c(V)$.

c) $\Rightarrow$ b). Let $\varepsilon \in \mathbb{R}$, $0 < \varepsilon < \text{dist}(K,\partial U)$, and let $\rho \in \mathbb{R}$, $\rho > 0$, $\rho < \varepsilon/4$. Then $\rho < 2\rho < \varepsilon/2 < \text{dist}(K,\partial U)$. $B_\rho(K) \subset \subset \bar{B}_{2\rho}(K) \subset U$, and $B_\rho(K)$ is an open subset of U containing K. Applying Lemma 8.2 to $f \in \mathcal{H}(U;F)$, to the compact set K, to $\bar{B}_{2\rho}(K)$ and to $\varepsilon/4$ we have

2) $$\sum_{m=0}^{\infty} (\varepsilon/4)^m \sup_{x\in \bar{B}_\rho(K)} \left\|\frac{1}{k!} \hat{d}^k f(x)\right\|$$

$$\leq \sum_{m=0}^{\infty} [2(\rho+\varepsilon/4)]^m \sup_{x\in K} \left\|\frac{1}{m!} \hat{d}^m f(x)\right\|$$

$$\leq \sum_{m=0}^{\infty} \varepsilon^m \sup_{x\in K} \left\|\frac{1}{m!} \hat{d}^m f(x)\right\| .$$

Now, by hypothesis, given the real number $\varepsilon/4 > 0$ and the open subset $V = B_\rho(K)$ of U containing K, there exists a real number $c(\varepsilon, V) > 0$, which in fact depends only on ε (since ρ depends on ε) and which we may accordingly denote by $c(\varepsilon)$, such that

$$3) \quad p(f) \leq c(\varepsilon) \sum_{m=0}^{\infty} (\varepsilon/4)^m \sup_{x \in B_\rho(K)} \left\|\frac{1}{m!}\hat{d}^m f(x)\right\|$$

for every $f \in \mathcal{H}(U;F)$. From 2) and 3) we obtain:

$$p(f) \leq c(\varepsilon) \sum_{m=0}^{\infty} \varepsilon^m \sup_{x \in K} \left\|\frac{1}{m!}\hat{d}^m f(x)\right\|$$

for every $f \in \mathcal{H}(U;F)$.

b) $\Rightarrow$ a). Let V be an open subset of U containing K, and let $\rho \in \mathbb{R}$, with $0 < \rho < \mathrm{dist}(K, \partial V)$. Then $\bar{B}_\rho(K) \subset V$ and so, by the Cauchy inequalities (Proposition 4.4), we have

$$4) \quad \sup_{x \in K} \left\|\frac{1}{m!}\hat{d}^m f(x)\right\| \leq \frac{1}{\rho^m} \sup_{x \in V} \|f(x)\|$$

for every $f \in \mathcal{H}(U;F)$, $m \in \mathbb{N}$. Choose $\varepsilon = \rho/2$. By hypothesis, there is a real number $c(\varepsilon) > 0$ such that

$$p(f) \leq c(\varepsilon) \sum_{m=0}^{\infty} \varepsilon^m \sup_{x \in K} \left\|\frac{1}{m!}\hat{d}^m f(x)\right\|$$

for every $f \in \mathcal{H}(U;F)$. Thus, from 4),

$$5) \quad p(f) \leq c(\varepsilon) \sum_{m=0}^{\infty} \left(\frac{\varepsilon}{\rho}\right)^m \sup_{x \in V} \|f(x)\| \leq 2c(\varepsilon) \sup_{x \in V} \|f(x)\|$$

for every $f \in \mathcal{H}(U;F)$. As ε depends on ρ, which in turn depends on V, we may take $C(V) = 2c(\varepsilon)$. Therefore p is ported by K. Q.E.D.

PROPOSITION 8.6 Let $\xi \in U$, and let K be a compact ξ-balanced subset of U. In order that a seminorm p on $\mathcal{H}(U;F)$ be ported by K it is necessary and sufficient that for every open subset V of U which contains K, there is a real number $c(V) > 0$ such that

$$p(f) \leq c(V) \sum_{m=0}^{\infty} \sup_{x\in V} \left\|\frac{1}{m!} \hat{d}^m f(\xi)(x-\xi)\right\|$$

for every $f \in \mathcal{H}(U;F)$.

PROOF: The condition is necessary: Suppose that p is ported by K, and that V is an open subset of U containing K. Let

$$W = \{x \in V : (1-\lambda)\xi + \lambda x \in V, \quad \lambda \in \mathbb{C}, \quad |\lambda| \leq 1\}.$$

Then W is the largest open ξ-balanced subset of V, and $K \subset W$. By Corollary 5.1 we have

$$(*) \qquad f(x) = \sum_{m=0}^{\infty} P_m(x-\xi)$$

for every $x \in W$ and $f \in \mathcal{H}(U;F)$, where $P_m = \frac{1}{m!} \hat{d}^m f(\xi)$, $m \in \mathbb{N}$. Since p is ported by K, there is a real number $C(W) > 0$ such that

$$p(f) \leq C(W) \sup_{x\in W} \|f(x)\|$$

for every $f \in \mathcal{H}(U;F)$. Using $(*)$, we have

$$p(f) \leq C(W) \sup_{x\in W} \sum_{m=0}^{\infty} \|P_m(x-\xi)\|$$
$$\leq C(W) \sum_{m=0}^{\infty} \sup_{x\in W} \|P_m(x-\xi)\|$$

for every $f \in \mathcal{H}(U;F)$. Since W depends only on V, we may set $c(V) = c(W)$ and we have $p(f) \le c(V) \sum_{m=0}^{\infty} \sup_{x\in V} \|P_m(x-\xi)\|$ for every $f \in \mathcal{H}(U;F)$.

The condition is sufficient: Let V be an open subset of U containing K. By Remark 5.1 there exists an open subset W of V containing K, and a real number $\rho > 1$, such that $(1-\lambda)\xi + \lambda x \in V$ for every $x \in W$ and $\lambda \in C$, $|\lambda| \le \rho$. Applying the Cauchy integral formula, Proposition 4.2, we have, for $f \in \mathcal{H}(U;F)$,

$$\left\|\frac{1}{m!}\hat{d}^m f(\xi)(x-\xi)\right\| = \left\|\frac{1}{2\pi i}\int_{|\lambda|=\rho} \frac{f[(1-\lambda)\xi+\lambda x]}{\lambda^{m+1}}\,d\lambda\right\|$$

$$\le \frac{1}{\rho^m} \sup_{|\lambda|=\rho} \|f[(1-\lambda)\xi+\lambda x]\|$$

$$\le \frac{1}{\rho^m} \sup_{t\in V} \|f(t)\|$$

for every $x \in W$, $m = 0,1,\ldots$. By hypothesis, there exists a real number $c(W) > 0$ such that for every $f \in \mathcal{H}(U;F)$,

$$p(f) \le c(W) \sum_{m=0}^{\infty} \sup_{x\in W} \left\|\frac{1}{m!}\hat{d}^m f(\xi)(x-\xi)\right\|$$

and so

$$p(f) \le c(W) \sum_{m=0}^{\infty} \frac{1}{\rho^m} \sup_{x\in V} \|f(x)\|$$

$$= \frac{c(W)\rho}{\rho-1} \sup_{x\in V} \|f(x)\|.$$

Taking $c(V) = \dfrac{c(W)\rho}{\rho-1}$, we have

$$p(f) \le c(V) \sup_{x\in V} \|f(x)\|$$

for every $f \in \mathcal{H}(U;F)$. Hence p is ported by K. Q.E.D.

PROPOSITION 8.7 If U is ξ-balanced, the Taylor series about ξ of every $f \in \mathcal{H}(U;F)$ converges to f in the topology τ_ω.

PROOF: Let $f \in \mathcal{H}(U;F)$, and let p be a seminorm on $\mathcal{H}(U;F)$ which is ported by the compact set $K \subset U$. Given $\varepsilon > 0$ the set $W = \{g \in \mathcal{H}(U;F) : p(g) < \varepsilon\}$ is a neighbourhood of zero for the topology τ_ω. By Proposition 5.2 there is an open subset V_f of U containing K such that the Taylor series of f about ξ converges uniformly to f in V_f. Let $C(V_f) > 0$ be such that

$$(*) \qquad p(g) \leq C(V_f) \sup_{x \in V_f} \|g(x)\| \quad \text{for every} \quad g \in \mathcal{H}(U;F).$$

Now there exists $m_o \in \mathbb{N}$ such that

$$\|(\tau_{m,f,\xi} - f)(x)\| \leq \varepsilon / C(V_f)$$

for every $m \geq m_o$ and $x \in V_f$. Applying (*) to $f - \tau_{m,f,\xi}$ we have

$$p(f-\tau_{m,f,\xi}) \leq C(V_f)\varepsilon / C(V_f) = \varepsilon \quad \text{if} \quad m \geq m_o;$$

thus $f-\tau_{m,f,\xi} \in W$ if $m \geq m_o$. Therefore $\tau_{m,f,\xi} \to f$ as $m \to \infty$ in the topology τ_ω, for every $f \in \mathcal{H}(U;F)$. Q.E.D.

We now introduce another topology on $\mathcal{H}(U;F)$.

DEFINITION 8.8 Let I be a countable cover of U by non-empty open subsets. We denote by $\mathcal{H}_I(U;F)$ the vector space of holomorphic mappings of U into F which are bounded on each open set of I.

The natural topology of $\mathcal{H}_I(U;F)$ is the separated locally convex topology defined by the seminorms

$$p_V : f \in \mathcal{H}_I(U;F) \mapsto p_V(f) = \sup_{x \in V} \|f(x)\|$$

where V ranges over I.

PROPOSITION 8.8 If F is a Banach space then $\mathcal{H}_I(U;F)$ with the natural topology is a Fréchet space.

PROOF: Since I is countable, and the natural topology is clearly separated, $\mathcal{H}_I(U;F)$ with this topology is metrizable, whether or not F is complete.

Let $\{f_m\}_{m \in \mathbb{N}}$ be a Cauchy sequence in $\mathcal{H}_I(U;F)$. Then given $\varepsilon > 0$ and $V \in I$ there exists $m_o \in \mathbb{N}$ such that if $m,n \in \mathbb{N}$, $m \geq m_o$ and $n \geq m_o$, then

$$(*) \qquad \sup_{x \in V} \|f_m(x) - f_n(x)\| < \varepsilon .$$

If K is a compact subset of U, a classical compactness argument shows that for every $\varepsilon > 0$ there exists $m_o \in \mathbb{N}$ such that

$$(**) \qquad \sup_{x \in K} \|f_m(x) - f_n(x)\| < \varepsilon \quad \text{if} \quad m \geq m_o \quad \text{and} \quad n \geq m_o .$$

Thus $\{f_m\}_{m \in \mathbb{N}}$ is a Cauchy sequence for the topology τ induced on $\mathcal{H}_I(U;F)$ by the compact-open topology on $C(U;F)$. Since F is complete, it follows from Proposition 4.6 that the sequence $\{f_m\}_{m \in \mathbb{N}}$ converges in the compact-open topology to a function $f \in \mathcal{H}(U;F)$. We must show that $f \in \mathcal{H}_I(U;F)$, and that $\{f_n\}_{m \in \mathbb{N}}$ converges to f in the natural topology.

It follows from (**) that $f(x) = \lim_{m \to \infty} f_m(x)$ for each $x \in V$, and so, by (*), if $V \in I$ and $\varepsilon > 0$, there exists

$m_o \in \mathbb{N}$ such that

$$(***) \qquad \sup_{x\in V} \|f(x)-f_n(x)\| \leq \varepsilon \quad \text{for} \quad n \geq m_o.$$

Therefore $\sup_{x\in V} \|f(x)\| \leq \sup_{x\in V} \|f_{m_o}(x)\| + \varepsilon < +\infty$, and so $f \in \mathcal{H}_I(U;F)$. Finally, (***) shows that $\{f_m\}_{m\in\mathbb{N}}$ converges to f in the natural topology of $\mathcal{H}_I(U;F)$. Q.E.D.

REMARK 8.6: a) Let I_1 and I_2 be countable covers of U by non-empty open subsets. I_2 is said to be finer than I_1 if for every $V_2 \in I_2$ there exists $V_1 \in I_1$ such that $V_2 \subset V_1$. If I_2 is finer than I_1 and if $f \in \mathcal{H}_{I_1}(U;F)$, then given $V_2 \in I_2$, there exists $V_1 \in I_1$ such that $V_2 \subset V_1$, and it follows that

$$\sup_{x\in V_2} \|f(x)\| \leq \sup_{x\in V_1} \|f(x)\|;$$

in other words, $f \in \mathcal{H}_{I_2}(U;F)$ and $p_{V_2}(f) \leq p_{V_1}(f)$. Therefore $\mathcal{H}_{I_1}(U;F) \subset \mathcal{H}_{I_2}(U;F)$, and the inclusion mapping is continuous for the respective natural topologies.

b) Let $I_1 = \{V_m : m \in \mathbb{N}\}$ and $I_2 = \{W_n : n \in \mathbb{N}\}$ be two countable covers of U by non-empty open subsets. Then $I_3 = \{V_m \cap W_n : m \in \mathbb{N},\ n \in \mathbb{N}$ and $V_m \cap W_n \neq \emptyset\}$ is also a countable cover of U by non-empty open subsets. Let $\mathfrak{R}$ be the set of all countable covers of U by open subsets, and for $I_1, I_2 \in \mathfrak{R}$ define $I_1 \leq I_2$ if I_2 is finer than I_1. Then $\leq$ is a filtered, or directed, partial order on $\mathfrak{R}$.

c) $\mathcal{H}(U;F) = \bigcup_{I\in\mathfrak{R}} \mathcal{H}_I(U;F)$; for if $f \in \mathcal{H}(U;F)$, let $V_k = \{x \in U : \|f(x)\| < k\}$ for $k \in \mathbb{N}$, $k \geq 1$. Then

$\{V_k\}_{k=1,2,\ldots}$ is an increasing sequence of open subsets of U whose union is U, and there exists $k_o \in \mathbb{N}$, $k_o \geq 1$ such that $V_k \neq \emptyset$ for $k \geq k_o$. Let $I = \{V_k\}_{k\in\mathbb{N}, k\geq k_o}$; then $I \in \mathfrak{R}$ and $f \in \mathcal{H}_I(U;F)$.

DEFINITION 8.9 τ_δ is the finest locally convex topology on $\mathcal{H}(U;F)$ for which the inclusion mappings $\mathcal{H}_I(U;F) \subset \mathcal{H}(U;F)$ are all continuous, where I ranges over the set $\mathfrak{R}$ of countable covers of U by non-empty open subsets of U and each $\mathcal{H}_I(U;F)$ carries its natural topology.

REMARK 8.8 The topology τ_δ was introduced independently and at the same time by Coeuré [22] in the separable case and by Nachbin [93] in the general case.

REMARK 8.9 The following are equivalent descriptions of τ_δ:

a) τ_δ is the final locally convex topology of the natural topologies of $\mathcal{H}_I(U;F)$ with respect to the inclusions $\mathcal{H}_I(U;F) \subset \mathcal{H}(U;F)$, as I ranges over $\mathfrak{R}$.

b) Given I_1 and I_2 belonging to $\mathfrak{R}$ with $I_1 \leq I_2$, denote by j_{I_2,I_1} the inclusion $\mathcal{H}_{I_1}(U;F) \subset \mathcal{H}_{I_2}(U;F)$. Then $\mathcal{H}(U;F)$ with the topology τ_δ is the inductive limit of the inductive system $(\mathcal{H}_I(U;F),\ j_{I,J})_{I,J\in\mathfrak{R}}$.

PROPOSITION 8.9 $\tau_\omega \leq \tau_\delta$.

PROOF: Let p be a seminorm on $\mathcal{H}(U;F)$ which is ported by a compact subset K of U. We prove that $p|_{\mathcal{H}_I(U;F)}$ is continuous for every $I \in \mathfrak{R}$, from which it follows that $\tau_\omega \leq \tau_\delta$. Fix $I \in \mathfrak{R}$. Then there exist elements $V_1,\ldots,V_k$ of

I such that $K \subset V_1 \cup \dots \cup V_k = V$, and since p is ported by K, there is a real number $C(V) > 0$ such that

$$(*) \qquad p(f) \leq C(V) \sup_{x \in V} \|f(x)\| \quad \text{for every } f \in \mathcal{H}(U;F).$$

Now, for every mapping f of U into F we have

$$\sup_{x \in V} \|f(x)\| = \sup_{1 \leq i \leq k} \{\sup_{x \in V_i} \|f(x)\|\},$$

and in particular, if $f \in \mathcal{H}_I(U;F)$,

$$\sup_{1 \leq i \leq k} \{\sup_{x \in V_i} \|f(x)\|\} = \sup_{1 \leq i \leq k} \{p_{V_i}(f)\}.$$

Therefore $(*)$ can be written:

$$p(f) \leq C(V) \sup_{1 \leq i \leq k} \{p_{V_i}(f)\} \quad \text{for every } f \in \mathcal{H}_I(U;F),$$

which shows that $p/\mathcal{H}_I(U;F)$ is continuous. Q.E.D.

REMARK 8.10: We have the following relation between the various topologies of $\mathcal{H}(U;F)$:

$$\tau_o \leq \tau_m \leq \tau_{m+1} \leq \tau_\infty \leq \tau_\omega \leq \tau_\delta \quad \text{for every } m \in \mathbb{N}.$$

PROPOSITION 8.10 Let $\chi \subset \mathcal{H}(U;F)$. The following are equivalent:

1) χ is bounded for the topology τ_δ.

2) χ is bounded for the topology τ_o.

PROOF: 1) $\Rightarrow$ 2) is trivial, since $\tau_o \leq \tau_\delta$.

2) $\Rightarrow$ 1). Let K be a compact subset of U, and let $p_K: \mathcal{H}(U;F) \to \mathbb{R}$ be the seminorm $p_K(f) = \sup_{x \in K} \|f(x)\|$, $f \in \mathcal{H}(U;F)$.

Then $p_K^{-1}([0,1]) = \{f \in \mathcal{H}(U;F) : p_K(f) \leq 1\}$ is a τ_o-neighbourhood of 0 in $\mathcal{H}(U;F)$ and so, since χ is τ_o-bounded, there exists $\rho > 0$ such that $\rho\chi \subset p_K^{-1}([0,1])$, that is,

$$\sup\{\|f(x)\| : x \in K, f \in \chi\} \leq \rho^{-1} < +\infty.$$

Hence χ is bounded on every compact subset of U, and it follows by methods analogous to those used in the proof of Lemma 6.1 that χ is locally bounded in U.

For $k \in \mathbb{N}$, $k \geq 1$, let $F_k = \{x \in U : \|f(x)\| \leq k$ for every $f \in \chi\}$ and let V_k be $\mathring{F}_k$, the interior of F_k. Then $\{V_k\}_{k=1,2,\ldots}$ is a countable cover of U by open subsets. To see this, let $x \in U$. Since χ is locally bounded, there exists an open neighbourhood W_x of x contained in U such that $\sup\{\|f(y)\| : f \in \chi, y \in W_x\} < +\infty$. It follows that W_x is contained in F_k, for some $k \in \mathbb{N}$, $k \geq 1$, and then since W_x is open, $W_x \subset \mathring{F}_k = V_k$, and so $x \in V_k$. Thus there exists $k_o \in \mathbb{N}$, $k_o \geq 1$, such that $V_k \neq \emptyset$ for $k \geq k_o$, and then $I = \{V_k : k \in \mathbb{N}, k \geq k_o\}$ is a countable cover of U by non-empty open subsets. For every $f \in \chi$ and every $k \in \mathbb{N}$, $k \geq k_o$, we have, by the definitions of V_k and F_k,

$$\sup_{x \in V_k} \|f(x)\| \leq \sup_{x \in F_k} \|f(x)\| \leq k;$$

in other words, $\chi \subset \mathcal{H}_I(U;F)$ and $p_{V_k}(f) \leq k$ for every $f \in \chi$, $k \in \mathbb{N}$, $k \geq k_o$. Hence χ is bounded in $\mathcal{H}_I(U;F)$ for the natural topology. Since the inclusion $\mathcal{H}_I(U;F) \subset \mathcal{H}(U;F)$ is continuous for the natural topology on $\mathcal{H}_I(U;F)$ and the topology τ_δ on $\mathcal{H}(U;F)$ we have that χ is τ_δ-bounded in $\mathcal{H}(U;F)$.

Q.E.D.

COROLLARY 8.1 The topologies τ_o, τ_m, τ_{m+1}, τ_∞, τ_ω and τ_δ $(m \in \mathbb{N})$ all have the same bounded sets.

REMARK 8.11: A locally convex topology τ on a vector space E is bornological if the following condition is satisfied:

For every locally convex topology τ' on E which has the same bounded sets as τ, $\tau' \leq \tau$.

It can be shown that for every locally convex topology τ on a vector space E there is a unique bornological topology τ_b on E such that τ and τ_b have the same bounded sets. τ_b is known as the bornological topology associated with τ. Thus $\tau \leq \tau_b$, and τ is bornological if and only if $\tau = \tau_b$.

PROPOSITION 8.11 The bornological topology associated with τ_o is τ_δ; that is, $\tau_\delta = (\tau_o)_b$.

PROOF: For every $I \in \mathfrak{R}$, the natural topology of $\mathcal{H}_I(U;F)$ is locally convex and metrizable, and hence is bornological. Therefore τ_δ, being the inductive limit of bornological topologies, is bornological. On the other hand, we know from Proposition 8.10 that τ_o and τ_δ have the same bounded sets. It follows by Remark 8.9 that τ_δ is the bornological topology associated with τ_o. Q.E.D.

COROLLARY 8.2 τ_δ is the bornological topology associated with τ_i, where $i \in \mathbb{N} \cup \{\infty\} \cup \{\omega\}$.

PROPOSITION 8.12 a) If the dimension of E is finite, or if $F = \{0\}$, then $\tau_o = \tau_m = \tau_{m+1} = \tau_\infty = \tau_\omega = \tau_\delta$ for every $m \in \mathbb{N}$.

b) If the dimension of E is infinite and $F \neq \{0\}$ then $\tau_o < \tau_m < \tau_{m+1} < \tau_\infty < \tau_\omega$ for every $m \in \mathbb{N}$.

PROOF: a) If $F = \{0\}$ then $\mathcal{H}(U;F) = \{0\}$ and the assertion is obvious. Suppose now that the dimension of E is finite and $F \neq \{0\}$. Since E is finite dimensional there exists a countable cover I of U by non-empty open subsets such that

1) The closure $\bar{V}$ of each $V \in I$ is compact and $\bar{V} \subset U$.

2) For every compact subset K of U these exists $V \in I$ such that $K \subset V$.

Then for every $f \in \mathcal{H}(U;F)$ and every $V \in I$,

$$\sup_{x \in V} \|f(x)\| = \sup_{x \in \bar{V}} \|f(x)\| < \infty,$$

since by construction $\bar{V}$ is compact. Hence, for every such cover I, $\mathcal{H}(U;F) = \mathcal{H}_I(U;F)$. It follows that τ_δ coincides with the natural topology of $\mathcal{H}_I(U;F)$, and since this natural topology coincides with τ_o for such an I, we have $\tau_\delta = \tau_o$.

b) Suppose that the dimension of E is infinite and $F \neq \{0\}$. We shall prove first that $\tau_{m-1} < \tau_m$ for every $m \in N$, $m \geq 1$.

Suppose that $\tau_{m-1} = \tau_m$ for some $m \in \mathbb{N}$, $m \geq 1$; if $\xi \in U$, the linear mapping

$$f \in \mathcal{H}(U;F) \mapsto \hat{d}^m f(\xi) \in \mathcal{P}(^m E;F)$$

is continuous when $\mathcal{H}(U;F)$ carries the topology τ_m, and hence is continuous for $\tau_{m-1} = \tau_m$. Therefore, there exists a real number $c > 0$ and a compact subset K of U, such that

$$(*) \qquad \left\|\frac{1}{m!}\hat{d}^m f(\xi)\right\| \leq c \sum_{i=0}^{m-1} \sup_{x\in K} \left\|\frac{1}{i!}\hat{d}^i f(x)\right\|$$

for every $f \in \mathcal{H}(U;F)$.

Choose $b \in F$, $b \neq 0$, and $\psi \in E'$, the topological dual of E, and consider the mapping $\psi^m b: E \to F$ defined by $(\psi^m b)(x) = [\psi(x)]^m \cdot b$, $x \in E$. Then $\psi^m b$ is a continuous m-homogeneous polynomial from E into F, being the restriction to the diagonal of E^m of the symmetric continuous m-linear mapping

$$(x_1,\ldots,x_m) \in E^m \to \psi(x_1)\ldots\psi(x_m)\cdot b \in F.$$

Let $f = \psi^m b/U$. Then

$$\frac{1}{i!}\hat{d}^i f(x) = \binom{m}{i} [\psi(x)]^{m-i}\cdot\psi^i b$$

for $i = 0,1,\ldots,m$, $x \in U$. In particular, $\frac{1}{m!}\hat{d}^m f(x) = \psi^m b$. Applying $(*)$ to f we obtain

$$\|\psi^m b\| \leq c \sum_{i=0}^{m-1} \binom{m}{i} \sup_{x\in K} |[\psi(x)]^{m-i}| \|\psi^i b\|$$

$$= c \sum_{i=0}^{m-1} \binom{m}{i} \|\psi\|^i \sup_{x\in K} |[\psi(x)]^{m-i}| \|b\|.$$

Since $\|b\| \neq 0$,

$$\|\psi^m\| \leq c \sum_{i=0}^{m-1} \binom{m}{i} \|\psi\|^i \sup_{x\in K} |\psi(x)|^{m-i}$$

which implies that

$$c\|\psi\|^m + \|\psi\|^m \leq c(\|\psi\| + \sup_{x\in K} |\psi(x)|)^m,$$

whence $\sqrt[m]{1+c}\, \|\psi\| \leq \sqrt[m]{c}\, (\|\psi\| + \sup_{x\in K} |\psi(x)|)$.

Therefore

$$\|\psi\| \leq \frac{\sqrt[m]{c}}{\sqrt[m]{1+c} - \sqrt[m]{c}} \sup_{x \in K} |\psi(x)|,$$

which can be expressed as:

$$\|\psi\| \leq \lambda \sup_{x \in K} |\psi(x)| = \sup_{x \in \lambda K} |\psi(x)|,$$

where

$$\lambda = \frac{\sqrt[m]{c}}{\sqrt[m]{1+c} - \sqrt[m]{c}}.$$

The set $M = \lambda K$ is compact; if we set $B = \{x \in E : \|x\| \leq 1\}$, then

$$|\psi(x)| \leq \|\psi\| \|x\| \leq \|\psi\| \leq \sup_{x \in M} |\psi(x)|$$

for every $x \in B$. Therefore

$$\sup_{x \in B} |\psi(x)| \leq \sup_{x \in M} |\psi(x)|.$$

Then $\psi \in M^o = \{\psi \in E' : |\psi(x)| \leq 1 \text{ for every } x \in M\}$ implies that $\psi \in B^o$; thus $M^o \subset B^o$, from which we deduce.

$$(**) \qquad B^{oo} \subset M^{oo}.$$

In the sequel we shall need the following results from the theory of topological vector spaces:

1) If E is a locally convex space and $A \subset E$, then A^{oo} is the closed convex balanced hull of A.

2) Riesz' Theorem: If the closed unit ball of a normed space E is precompact, then E is finite dimensional.

3) If M is a compact subset of a locally convex space, then the closed convex balanced hull of M is precompact.

Applying 1) to (**) we find that $B = B^{oo} \subset M^{oo}$, since B is closed, convex and balanced. By 3) M^{oo} is precompact, and since a subset of a precompact set is precompact, B is precompact. But then by 2) the dimension of E must be finite, contradicting our initial hypothesis.

We deduce from this that $\tau_m < \tau_\infty$ for every $m \in N$, for if $\tau_{m_o} = \tau_\infty$ for some $m_o \in \mathbb{N}$, we would have $\tau_{m_o} = \tau_{m_o+1}$, since $\tau_{m_o} \leq \tau_{m_o+1} \leq \tau_\infty$.

Finally, we prove that $\tau_\infty < \tau_\omega$. Assuming the contrary, let $\alpha = \{\alpha_m\}_{m\in\mathbb{N}}$ be a sequence of positive real numbers for which $\lim_{m\to\infty} \alpha_m^{1/m} = 0$, let $\xi \in U$, and let p be the seminorm

$$f \in \mathcal{H}(U;F) \mapsto p(f) = \sum_{m=0}^{\infty} \alpha_m \left\|\frac{1}{m!} \hat{d}^m f(\xi)\right\| \in \mathbb{R}.$$

Example 8.3 shows that p is ported by the compact set $\{\xi\}$. Therefore, if $\tau_\omega = \tau_\infty$, there exists a compact set $K \subset U$, a real number $c > 0$, and $m \in \mathbb{N}$, $m \geq 1$, such that

$$(*) \qquad p(f) \leq c \sum_{i=0}^{m-1} \sup_{x\in K} \left\|\frac{1}{i!} \hat{d}^i f(x)\right\|$$

for every $f \in \mathcal{H}(U;F)$. From the definition of p we have

$$(**) \qquad \left\|\frac{1}{m!} \hat{d}^m f(\xi)\right\| \leq \frac{1}{\alpha_m} p(f) \quad \text{for every } f \in \mathcal{H}(U;F).$$

Combining (*) and (**),

$$\left\|\frac{1}{m!} \hat{d}^m f(\xi)\right\| \leq \frac{c}{\alpha_m} \sum_{i=0}^{m-1} \sup_{x\in K} \left\|\frac{1}{i!} \hat{d}^i f(x)\right\|$$

for every $f \in \mathcal{H}(U;F)$.

Proceeding as before, with $f = \psi^m b$, we obtain a contradiction. Therefore $\tau_\infty < \tau_\omega$. Q.E.D.

REMARK 8.12: In the case in which E is infinite dimensional and $F \neq \{0\}$ it can happen either that $\tau_\omega = \tau_\delta$ or that $\tau_\omega < \tau_\delta$.

To illustrate this, we first recall some definitions:

A sequence $\{x_n\}_{n=1,2,...}$ in a Banach space E is a Schauder basis for E if every $x \in E$ can be written uniquely in the form $x = \sum_{n=1}^{\infty} \alpha_n x_n$, $\alpha_n \in \mathbb{C}$ for each $n \in \mathbb{N}$.

A Schauder basis $\{x_n\}_{n=1,2,...}$ is unconditional when every convergent series of the form $\sum_{n=1}^{\infty} \alpha_n x_n$ is unconditionally convergent, that is, for every permutation π of $\mathbb{N}^* = \{1,2,...\}$, $\sum_{n=1}^{\infty} \alpha_{\pi(n)} x_{\pi(n)} = \sum_{n=1}^{\infty} \alpha_n x_n$.

The following result is due to Dineen [32]:

Let E and F be complex Banach spaces, U a non-empty open subset of E, and $\xi \in U$. If U is ξ-balanced and if E has an unconditional Schauder basis, then the topologies τ_ω and τ_δ coincide on $\mathcal{H}(U;F)$.

REMARK 8.13: Since every Banach space with an unconditional basis is separable, it is natural to ask the following question:

Let E and F be Banach spaces, and U a non-empty open subset of E. If E has an unconditional basis, or, more generally, if E is separable, do the topologies τ_ω and τ_δ coincide on $\mathcal{H}(U;F)$, that is, is τ_ω bornological?

REMARK 8.14: The spaces c_o and ℓ^p $(p \in \mathbb{R},\ p \geq 1)$ have unconditional Schauder basis; $C[0,1]$ does not have an unconditional Schauder basis. The space $L^p[0,1]$ $(p \in \mathbb{R}, p \geq 1)$

has an unconditional Schauder basis but, in general, $L^p(\Omega,\mathfrak{F},\mu)$ does not have an unconditional Schauder basis.

EXAMPLE 8.4 If $E = \ell^\infty$, then on $\mathcal{H}(E) = \mathcal{H}(E;\mathbb{C})$, the topologies τ_ω and τ_δ do not coincide, that is, $\tau_\omega < \tau_\delta$. See Dineen [31].

PROPOSITION 8.13 $\mathcal{H}(E;\mathbb{C})$ is τ_ω-complete; if U is ξ-balanced for some $\xi \in U$, then $\mathcal{H}(U;\mathbb{C})$ is τ_ω-complete. (See Dineen [30]).

CHAPTER 9

UNIQUENESS OF ANALYTIC CONTINUATION

PROPOSITION 9.1 Let U be a non-empty open connected set and $f \in \mathcal{H}(U;F)$. The following are equivalent:

a) f is identically zero in U.

b) There exists a non-empty open set $V \subset U$ such that $f\big|_V = 0$.

c) There exists a point $\xi \in U$ such that $\frac{1}{k!} d^k f(\xi) = 0$ for every $k \in \mathbb{N}$.

d) There exists a point $\xi \in U$ such that $\frac{1}{k!} \hat{d}^k f(\xi) = 0$ for every $k \in \mathbb{N}$.

PROOF: The implications a) $\Rightarrow$ b) $\Rightarrow$ c) $\Rightarrow$ d) are trivial.

d) $\Rightarrow$ a). Suppose that $\frac{1}{k!} \hat{d}^k f(\xi) = 0$ for every $k \in \mathbb{N}$. Let

$$S = \{x \in U : \frac{1}{k!} \hat{d}^k f(x) = 0 \text{ for every } k \in \mathbb{N}\}.$$

Then $\xi \in S$, so S is not empty.

S is open in U: if $x \in S$, and R_x is the radius of convergence of the Taylor series of f at x then $f(y) = 0$ for every $y \in B_{R_x}(x)$ and so by Proposition 2.2, $\frac{1}{k!} \hat{d}^k f(y) = 0$ for every $k \in \mathbb{N}$. Thus $y \in S$.

S is closed in U: if x does not lie in S, there is at least one index $k_o \in \mathbb{N}$ such that $\frac{1}{k_o!} \hat{d}^{k_o} f(x) \neq 0$.

Consider the mapping

$$\frac{1}{k_o!}\hat{d}^{k_o}f\colon y \in U \mapsto \frac{1}{k_o!}\hat{d}^{k_o}f(y) \in \mathcal{P}(^{k_o}E;F).$$

By Proposition 5.4 this mapping is holomorphic. In particular, it is continuous, and so there exists a neighbourhood V_x of x in U on which it does not vanish. Thus $\frac{1}{k_o!}\hat{d}^{k_o}f(y) \neq 0$ for every $y \in V_x$, and so $V_x \subset U\backslash S$, which shows that S is closed in U.

Since U is connected, $S = U$, and so f is identically zero in U. Q.E.D.

PROPOSITION 9.2 Let U be a non-empty open connected set and $f \in \mathcal{H}(U;F)$. Then the following sets all generate the same closed subspace of F:

a) $f(U)$, the image of U by f.

b) $f(V)$, the image of a non-empty open subset V of U by f.

c) $A = \{\frac{1}{m!} d^m f(\xi)(x_1,\dots,x_m) : m \in \mathbb{N}, (x_1,\dots,x_m) \in E^m\}$ where ξ is any point of U.

d) $B = \{\frac{1}{m!}\hat{d}^m f(\xi)(x) : m \in \mathbb{N}, x \in E\}$ where ξ is any point of U.

PROOF: Denoting by F_U and F_V the closed subspaces of F generated by $f(U)$ and $f(V)$ respectively, we have $F_V \subset F_U$, since $V \subset U$. Let $\pi_U\colon F \to F/F_U$ and $\pi_V\colon F \to F/F_V$ be the canonical mappings. Since π_V is linear and continuous, by Proposition 3.2 $\pi_V \circ f \in \mathcal{H}(U;F/F_V)$, and $\hat{d}^m(\pi_V \circ f)(\xi) = \pi_V \circ \hat{d}^m f(\xi)$ for every $\xi \in U$, $m \in \mathbb{N}$. Since $f(V) \subset F_V$,

$\pi_V \circ f$ vanishes on V and so by Proposition 9.1 $\pi_V \circ f$ vanishes identically on U. This implies that $f(U) \subset F_V$, and hence $F_U \subset F_V$.

Now let F_A and F_B denote the closed subspaces generated by A and B respectively: it is easy to see that $F_B \subset F_A$. On the other hand, the polarization formula shows that every vector in A is a linear combination of vectors in B. Thus $F_B = F_A$.

Finally, we show that $F_U = F_B$. By Proposition 3.2 $\pi_U \circ f \in \mathcal{H}(U;F/F_U)$ and $\hat{d}^m(\pi_U \circ f)(\xi) = \pi_U \circ \hat{d}^m f(\xi)$. Since $f(U) \subset F_U$, $\pi_U \circ f$ is identically zero in U, so that $\hat{d}^m(\pi_U \circ f)(\xi) = 0$ for every $m \in \mathbb{N}$. Therefore $\pi_U(\frac{1}{m!} \hat{d}^m f(\xi)) = 0$ for every $m \in \mathbb{N}$, which shows that $\frac{1}{m!} \hat{d}^m f(\xi)(x) \in F_U$ for every $m \in \mathbb{N}$, $x \in E$. Hence $F_B \subset F_U$.

On the other hand, since $\frac{1}{m!} \hat{d}^m f(\xi)(x) \in F_B$ for every $m \in \mathbb{N}$, $x \in E$, $\pi_B(\frac{1}{m!} \hat{d}^m f(\xi)(x)) = 0$, and so $\frac{1}{m!} \hat{d}^m(\pi_B \circ f)(\xi)(x) = 0$ for every $m \in \mathbb{N}$, $x \in E$. Now by Proposition 9.1, $\pi_B \circ f = 0$ in U, which implies that $f(U) \subset F_B$; therefore $F_U \subset F_B$. Q.E.D.

REMARK 9.1 It is easy to show that Propositions 9.1 and 9.2 are equivalent.

CHAPTER 10

THE MAXIMUM PRINCIPLE

PROPOSITION 10.1 (The maximum modulus theorem) Let U be an open connected set and $f \in \mathcal{H}(U;\mathbb{C})$. If $\xi \in U$ is such that $|f(\xi)| \geq |f(x)|$ for every $x \in U$, then f is constant on U.

PROOF: We assume known the case $E = \mathbb{C}$. Let E be any normed space, and r a positive real number such that $B_r(\xi) \subset U$. For $x \in B_r(\xi)$, $x \neq \xi$,

$$\rho = \frac{r}{\|x-\xi\|} > 1 \text{ and } (1-\lambda)\xi+\lambda x \in U \text{ for } \lambda \in D = \{\lambda \in \mathbb{C} : |\lambda| < \rho\}.$$

We define a function

$$g: \lambda \in D \mapsto g(\lambda) = f[(1-\lambda)\xi+\lambda x] \in \mathbb{C};$$

then $g \in \mathcal{H}(D;\mathbb{C})$ and by hypothesis $|g(0)| \geq |g(\lambda)|$ for every $\lambda \in D$. Therefore g is constant on D, and since $1 \in D$, we have $g(0) = g(1)$, that is, $f(x) = f(\xi)$.

Since x was an arbitrary element of $B_r(\xi)$, $x \neq \xi$, f is constant on $B_r(\xi)$. By Proposition 9.1, since U is connected, f is constant in U. Q.E.D.

REMARK 10.1: The following simple example shows that Proposition 10.1 is false in general for vector valued functions.

EXAMPLE 10.1 Let $E = \mathbb{C}$ and let F be $\mathbb{C}^3$ with the maximum norm. Thus, for $y = (y_1,y_2,y_3) \in F$, $\|y\| =$

$= \max\{|y_1|,|y_2|,|y_3|\}$. Let $U = \{x \in \mathbb{C} : |x| < 1\}$, and define $f: U \to F$ by $f(x) = (1,x,x^2)$, $x \in U$. Then U is connected, f is holomorphic in U, and $\|f(x)\| = \max\{1,|x|,|x|^2\} = 1$ for every $x \in U$. In particular, $\|f(0)\| \geq \|f(x)\|$ for every $x \in U$, so that hypothesis analogous to those of Proposition 10.1 are satisfied, while f is not constant in U.

REMARK 10.2 For further information on this question we refer the reader to the works of Thorp and Whitley [137] and of Vasentini [139].

PROPOSITION 10.2 (The maximum norm theorem) Let U be a bounded open subset of E, and $f \in C(\bar{U};F)$ such that $f|_U \in \mathcal{H}(U;F)$. Then

$$\sup_{x\in\bar{U}} \|f(x)\| = \sup_{x\in U} \|f(x)\| = \sup_{x\in\partial U} \|f(x)\|.$$

PROOF: We note first that the equation

$$\sup_{x\in\bar{U}} \|f(x)\| = \sup_{x\in U} \|f(x)\|$$

follows from the continuity of f in $\bar{U}$. And since $\partial U \subset \bar{U}$, we have

$$\sup_{x\in\partial U} \|f(x)\| \leq \sup_{x\in\bar{U}} \|f(x)\|.$$

Thus we must prove that

$$\sup_{x\in\partial U} \|f(x)\| \geq \sup_{x\in\bar{U}} \|f(x)\|.$$

We shall divide the proof into three parts.

a) $E = F = \mathbb{C}$: we assume this case as known.

b) $E = \mathbb{C}$, F an arbitrary normed space, $F \neq \{0\}$. We can reduce this to case a) by using the Hahn-Banach theorem. Fixing $x \in U$, this theorem yields a $\psi \in F'$ such that $\|\psi\| = 1$ and $\psi[f(x)] = \|f(x)\|$. Now, by Proposition 3.2, $\psi \circ f \in \mathcal{H}(U;\mathbb{C})$ and so by a),

$$(*) \qquad |\psi[f(x)]| \leq \sup_{t \in \partial U} |\psi[f(t)]|.$$

But $|\psi[f(t)]| \leq \|\psi\| \cdot \|f(t)\| = \|f(t)\|$ and $\psi[f(x)] = \|f(x)\|$, and so (*) implies that

$$\|f(x)\| \leq \sup_{t \in \partial U} \|f(t)\|.$$

Since x was an arbitrary element of U this completes the proof in case b).

c) E and F are normed spaces, $F \neq \{0\}$. We leave it to the reader to show that this can be reduced to case b).

Finally, the case $F = \{0\}$ is trivial. Q.E.D.

REMARK 10.3 The following simple example shows that the hypothesis of boundedness of U is essential, even when the dimension of E is finite.

EXAMPLE 10.2 Let $E = F = \mathbb{C}$, $U = \{z \in \mathbb{C} : \operatorname{Re} z > 0\}$, and let $f: \mathbb{C} \to \mathbb{C}$ be $f(z) = e^z$. Then $f \in \mathcal{H}(\mathbb{C};\mathbb{C})$, and so $f \in C(\bar{U};\mathbb{C})$, and $f \in \mathcal{H}(U;\mathbb{C})$. However, since $\partial U = \{ix : x \in \mathbb{R}\}$, we have

$$\sup_{z \in \partial U} |f(z)| = 1 < \sup_{z \in U} |f(z)| = +\infty.$$

CHAPTER 11

HOLOMORPHIC MAPPINGS OF BOUNDED TYPE

DEFINITION 11.1 A mapping $f: E \to F$ is said to be of bounded type if f is bounded on every bounded subset of E. $\mathcal{H}_b(E;F)$ denotes the subspace of $\mathcal{H}(E;F)$ consisting of the entire mappings of bounded type.

When E is finite dimensional, or $F = \{0\}$, every entire mapping of E into F is of bounded type. Examples 5.1, 5.2 and 5.3 show that there exist entire mappings which are not of bounded type.

PROPOSITION 11.1 Let F be complete. For $f \in \mathcal{H}(E;F)$, the following are equivalent:

a) f is of bounded type.

b) There exists $\xi \in E$ such that $\lim_{m\to\infty} \|\frac{1}{m!} \hat{d}^m f(\xi)\|^{1/m} = 0.$

c) $\lim_{m\to\infty} \|\frac{1}{m!} \hat{d}^m f(x)\|^{1/m} = 0$ for every $x \in E$.

d) There exists $\xi \in E$ such that $\lim_{m\to\infty} \|\frac{1}{m!} d^m f(\xi)\|^{1/m} = 0.$

e) $\lim_{m\to\infty} \|\frac{1}{m!} d^m f(x)\|^{1/m} = 0$ for every $x \in E$.

f) For every $x \in E$ the radius of convergence of the Taylor series of f at x is infinite.

PROOF: For $x \in E$, let $r_b(x)$ be the radius of boundedness of f at x, and let $R(x)$ be the radius of convergence of the Taylor series of f at x. Then condition a), which states that $r_b(x) = +\infty$ for every $x \in E$, is equivalent by Proposition 5.3, to the statement $R(x) = +\infty$ for every $x \in E$. Therefore a) $\Leftrightarrow$ f).

The Cauchy-Hadamard formula (Proposition 2.1) shows that f) is equivalent to c), and so a) $\Leftrightarrow$ f) $\Leftrightarrow$ c).

By Proposition 1.3,

$$\left\|\frac{1}{m!}\hat{d}^m f(x)\right\|^{1/m} \leq \left\|\frac{1}{m!} d^m f(x)\right\|^{1/m} \leq \left(\frac{m^m}{m!}\right)^{1/m} \left\|\frac{1}{m!}\hat{d}^m f(x)\right\|^{1/m}$$

for every $x \in E$ and $m \in \mathbb{N}$. Since $\lim_{m\to\infty} \left(\frac{m^m}{m!}\right)^{1/m} = e$, we have c) $\Leftrightarrow$ e) and b) $\Leftrightarrow$ d). Clearly, c) $\Rightarrow$ d), and so the proof is complete if we show that b) $\Rightarrow$ a).

The Cauchy-Hadamard formula (Proposition 2.1) applied to b) shows that $R(\xi) = +\infty$, and thus $r_b(\xi) = +\infty$. Now every bounded subset X of E is contained in a ball $B_\rho(\xi)$; therefore $f(X)$ is bounded for every bounded $X \subset E$. Q.E.D.

REMARK 11.1: If $\{a_m\}_{m\in\mathbb{N}}$ is a sequence of complex numbers, and $\xi \in \mathbb{C}$, the power series $\sum_{m=0}^{\infty} a_m(z-\xi)^m$ is the Taylor series at ξ of a function $f \in \mathcal{H}(\mathbb{C};\mathbb{C})$ if and only if $\lim_{m\to\infty} |a_m|^{1/m} = 0$. We shall see that in the general case the situation is not quite the same.

PROPOSITION 11.2 Let F be a Banach space, $\{A_m\}_{m\in\mathbb{N}}$ a sequence with $A_m \in \mathcal{L}_s(^mE;F)$, and $\xi \in E$. Then the power series $\sum_{m=0}^{\infty} A_m(x-\xi)^m$ about ξ is the Taylor series of a holomorphic mapping of bounded type from E into F if and only if $\lim_{m\to\infty} \|A_m\|^{1/m} = 0$.

PROOF: Suppose that $\sum_{m=0}^{\infty} A_m(x-\xi)^m$ is the Taylor series at ξ of an entire mapping of bounded type. Then the implication a) ⇒ e) of Proposition 11.1 and the uniqueness of the Taylor series shows that $\lim_{m\to\infty} \|A_m\|^{1/m} = 0$.

Conversely, suppose that $\lim_{m\to\infty} \|A_m\|^{1/m} = 0$. Given $\rho > 0$, there exists $m_o \in \mathbb{N}$, $m_o \geq 1$, such that $\|A_m\|^{1/m} \leq 1/2\rho$ for $m > m_o$. Then if $x \in B_\rho(\xi)$,

$$\sum_{m=0}^{\infty} \|A_m(x-\xi)^m\| \leq \sum_{m=0}^{m_o} \|A_m\| \|x-\xi\|^m + \sum_{m>m_o} \|A_m\| \|x-\xi\|^m$$

$$\leq \sum_{m=0}^{m_o} \|A_m\| \rho^m + \sum_{m>m_o} \frac{1}{(2\rho)^m} \rho^m$$

$$= \sum_{m=0}^{m_o} \|A_m\| \rho^m + \sum_{m>m_o} 2^{-m} < \infty .$$

Therefore, since F is complete, the series $\sum_{m=0}^{\infty} A_m(x-\xi)^m$ converges uniformly in every ball $B_\rho(\xi)$. In particular, this series converges uniformly in every bounded subset of E. Define

$$f(x) = \sum_{m=0}^{\infty} A_m(x-\xi)^m, \qquad x \in E.$$

If $g_\nu(x) = \sum_{m=0}^{\nu} A_m(x-\xi)^m$, $\nu \in \mathbb{N}$, then since compact sets are bounded, the sequence $\{g_\nu\}_{\nu\in\mathbb{N}}$ converges uniformly on every compact subset of E. Since $g_\nu \in \mathcal{H}(E;F)$ for every $\nu \in \mathbb{N}$, $f \in \overline{\mathcal{H}(E;F)}$, the closure being taken in $C(E;F)$ with the compact-open topology. By Proposition 4.6, since F is complete, $\mathcal{H}(E;F)$ is complete in the compact-open topology. Therefore $f \in \mathcal{H}(E;F)$ and, by the uniqueness of the Taylor series, $A_m = \frac{1}{m!} d^m f(\xi)$ for every $m \in \mathbb{N}$. The implication d) ⇒ a) of Proposition 11.1 shows that f is of bounded type.

Q.E.D.

We state the following results without proof:

PROPOSITION A: Let E be a normed space. The weak topology $\sigma(E',E)$ on E' and the topology of uniform convergence on compact subsets of E induce the same topology on each bounded subset of E'.

PROPOSITION B (The Josefson-Nissenzweig Theorem): Let E be a real or complex normed space of infinite dimension. Then there exists a sequence $\{\psi_m\}_{m\in\mathbb{N}} \subset E'$ such that

1) $\|\psi_m\| = 1$ for every $m \in \mathbb{N}$.

2) $\lim_{m\to\infty} \psi_m(x) = 0$ for every $x \in E$.

REMARK 11.2: Proposition B provided the solution to a long-standing conjecture. The proof can be found in Josefson [63] and Nissenzweig [112].

We shall use Propositions A and B to prove

PROPOSITION 11.3 The following statements are equivalent:

1) $\mathcal{H}_b(E;F) \neq \mathcal{H}(E;F)$.

2) The dimension of E is infinite and $F \neq \{0\}$.

PROOF: 1) $\Rightarrow$ 2) is clear, since if $F = \{0\}$ or if the dimension of E is finite, $\mathcal{H}_b(E;F) = \mathcal{H}(E;F)$.

2) $\Rightarrow$ 1). If E has infinite dimension, then by Proposition B there is a sequence $\{\psi_m\}_{m\in\mathbb{N}} \subset E'$ such that $\psi_m \to 0$ as $m \to \infty$ in the weak topology $\sigma(E',E)$, and $\|\psi_m\| = 1$ for every $m \in \mathbb{N}$. For each $x \in E$, since $\lim_{m\to\infty} \psi_m(x) = 0$, there exists $m_o \in \mathbb{N}$ such that $|\psi_m(x)| < 1/2$ for $m \geq m_o$. There-

fore the series $\sum_{m=0}^{\infty} [\psi_m(x)]^m$ is absolutely convergent for every $x \in E$, and we may define a mapping $f: E \to \mathbb{C}$ by

$$f(x) = \sum_{m=0}^{\infty} [\psi_m(x)]^m, \qquad x \in E.$$

We claim first that the series $\sum_{m=0}^{\infty} [\psi_m(x)]^m$ is uniformly convergent on the compact subsets of E. To see this, we note that the set $B = \{\psi_m : m \in \mathbb{N}\} \cup \{0\}$ is bounded in E', since $\|\psi_m\| = 1$ for every $m \in \mathbb{N}$. By Proposition A, the weak topology $\sigma(E',E)$ and the topology of uniform convergence on compact subsets of E must then agree on B, and since $\psi_m \to 0$ as $m \to \infty$ in the weak topology, it follows that $\psi_m \to 0$ as $m \to \infty$ in the topology of uniform convergence on compact subsets of E. Thus if K is a compact subset of E, there exists $m_1 \in \mathbb{N}$ such that $\sup_{x \in K} |\psi_m(x)| < 1/2$ if $m \in \mathbb{N}$, $m \geq m_1$. Then $|\psi_m(x)|^m < 1/2^m$ for every $x \in K$, $m \geq m_1$, and so the series $\sum_{m=0}^{\infty} (\psi_m)^m$ converges uniformly on K. Therefore $\sum_{m=0}^{\infty} (\psi_m)^m$ converges uniformly to f on the compact subsets of E.

Since E is metrizable, $C(E,\mathbb{C})$ is complete for the topology of uniform convergence on compact subsets of E. By Proposition 4.6, $\mathcal{H}(E;\mathbb{C})$ is a closed subspace of $C(E,\mathbb{C})$ for this topology, and hence is complete. But the partial sums $g_m = \sum_{k=0}^{m} (\psi_k)^k$ are holomorphic mappings of E into $\mathbb{C}$, and since $g_m \to f$ as $m \to \infty$ in the topology of uniform convergence on compact subsets of E, it follows that $f \in \mathcal{H}(E;\mathbb{C})$.

We now claim that f is not of bounded type. By virtue

of Proposition 11.1 it suffices to prove that $\|\frac{1}{m!}\hat{d}^m f(0)\|^{1/m}$ does not converge to 0 as m tends to infinity.

For every $m \in \mathbb{N}$ the mapping

$$\psi_m^m: x \in E \mapsto [\psi_m(x)]^m \in \mathbb{C}$$

is a continuous m-homogeneous polynomial from E into $\mathbb{C}$. It is easy to see that the series $\sum_{m=0}^{\infty} \psi_m^m$ converges uniformly in $\bar{B}_{1/2}(0)$, for if $x \in \bar{B}_{1/2}(0)$, then

$$\sum_{m=0}^{\infty} |[\psi_m(x)]^m| \leq \sum_{m=0}^{\infty} \|\psi_m\| \|x\|^m = \sum_{m=0}^{\infty} \|x\|^m \leq \sum_{m=0}^{\infty} (1/2)^m < \infty.$$

It follows from the definition of f and the uniqueness of the Taylor series that $\sum_{m=0}^{\infty} \psi_m^m$ is the Taylor series of f at 0; thus

$$\frac{1}{m!} \hat{d}^m f(0) = \psi_m^m \quad \text{for every } m \in \mathbb{N}.$$

Hence $\|\frac{1}{m!} \hat{d}^m f(0)\| = \|\psi_m\|^m = 1$ for every $m \in \mathbb{N}$, and so

$$\lim_{m\to\infty} \|\frac{1}{m!} \hat{d}^m f(0)\|^{1/m} = 1 \neq 0.$$

Therefore $f \notin \mathcal{H}_b(E;\mathbb{C})$.

Now if $F \neq \{0\}$, let $b \in F$, $b \neq 0$, and consider the mapping $f \cdot b: x \in E \mapsto f \cdot b(x) = f(x)b \in F$. Then f is entire, but is not of bounded type, since if X is a bounded subset of E for which f(X) is not bounded, it is clear that $f \cdot b$ is not bounded on X. This concludes the proof of 2) $\Rightarrow$ 1).

Q.E.D.

REMARK 11.3: In Definition 8.2 we have defined $f \in \mathcal{H}(U;F)$ to be of bounded type if f is bounded on every subset X of U such that X is bounded and $\mathrm{dist}(X,\partial U) > 0$. $\mathcal{H}_b(U;F)$ denotes the subspace of $\mathcal{H}(U;F)$ of mappings $f \in \mathcal{H}(U;F)$ which are of bounded type. We note that when $U = E$ this definition agrees with Definition 11.1, since every bounded subset of E is a positive distance from the boundary of E.

PROPOSITION 11.4 The following statements are equivalent:

a) $\mathcal{H}_b(U;F) \neq \mathcal{H}(U;F)$.

b) The dimension of E is infinite and $F \neq \{0\}$.

PROOF: a) $\Rightarrow$ b) is clear, since if the dimension of E is finite or if $F = \{0\}$, then $\mathcal{H}_b(U;F) = \mathcal{H}(U;F)$.

b) $\Rightarrow$ a). By Proposition 11.2 there exists $f \in \mathcal{H}(E;F)$ such that $f \notin \mathcal{H}_b(E;F)$. Let X be a bounded subset of E on which f is not bounded. Let $M = \sup\{\|x\| : x \in X\}$ and let $\xi \in U$, $\sigma \in R$, $\sigma > 0$, such that $\bar{B}_\sigma(\xi) \subset U$. Since X is bounded there exists a real number $\rho > 0$ such that $X \subset B_\rho(\xi)$, and we may assume that $M/\rho < 1$. Then the set $X' = \sigma\rho^{-1}X + \xi = \{\sigma\rho^{-1}x + \xi : x \in X\}$ is a subset of U which is bounded. Moreover X' is U-bounded. In fact, for every $x \in X$ we have

$$\|\sigma\rho^{-1}x + \xi - \xi\| = \|\sigma\rho^{-1}x\| \leq \sigma\frac{M}{\rho} = r < \sigma.$$

Therefore $X' \subset B_r(\xi)$ and since $\bar{B}_\sigma(\xi) \subset U$ it follows that $\mathrm{dist}(X',\partial U) > 0$.

Let g be the restriction to U of the composition $f \circ h \circ t$, where

$$t: x \in E \mapsto t(x) = x - \xi \in E$$

and

$$h: x \in E \mapsto h(x) = \sigma^{-1}\rho x \in E.$$

Thus $g(x) = f[\sigma^{-1}\rho(x-\xi)]$, $x \in U$. It is easy to see that $g \in \mathcal{H}(U;F)$, and since $h[t(X')] = X$, $g(X') = f(X)$. Hence $g \notin \mathcal{H}_b(U;F)$.

REMARK 11.4: In many problems concerning infinite dimensional normed spaces the space $\mathcal{H}_b(U;F)$ plays a role similar to that of $\mathcal{H}(U;F)$ in the case of finite dimensional spaces. One example of this is the study of domains of holomorphy and the Cartan-Thullen theorem.

CHAPTER 12

DOMAINS OF $\mathcal{H}_b$-HOLOMORPHY

We wish now to consider the following problem:

Given a non-empty open subset U of E, does there exist an open subset V of E, with $V \supsetneq U$, such that every $f \in \mathcal{H}(U;F)$ extends to a mapping $f \in \mathcal{H}(V;F)$? (If U and V are connected, Proposition 9.1 shows that this extension is unique). Equivalently, given a non-empty open subset U of E, does there exist an open subset V of E, with $V \supsetneq U$, such that the linear restriction mapping

$$f \in \mathcal{H}(V;F) \mapsto f/U \in \mathcal{H}(U;F)$$

is surjective?

It is easy to see that, without any further conditions, this question has a negative answer, even when the dimension of E is 1. For, if we take $E = F = \mathbb{C}$, $\xi \in E$ and $U = C-\{\xi\}$, the function

$$f: x \in U \mapsto f(x) = \frac{1}{x-\xi} \in \mathbb{C}$$

is holomorphic in U and has no holomorphic extension to $\mathbb{C}$.

In the general case, this problem is vary difficult, and there are few satisfactory results. We shall simplify the problem by substituting $\mathcal{H}_b(U;F)$ for $\mathcal{H}(U;F)$. More precisely, we shall characterise a large class of non-empty open subsets U of E for which there is no open set V, $V \supsetneq U$, such that the linear mapping

$$f \in \mathcal{H}_b(V;F) \to f/U \in \mathcal{H}(U;F)$$

is surjective.

We shall restrict our attention to the case $F = \mathbb{C}$. Thus most of the definitions and results which follow will refer to the space $\mathcal{H}_b(U;\mathbb{C})$, which we abbreviate by $\mathcal{H}_b(U)$.

REMARK 12.1: For the problem of holomorphic extension which we have posed, the case in which the dimension of E is 2 or more is totally different from the one-dimensional case.

PROPOSITION 12.1 Let E be a complex normed space of dimension greater than or equal to 2, and let F be a complex Banach space. If V is a non-empty open subset of E, $\xi \in V$ and $U = V-\{\xi\}$, then every $f \in \mathcal{H}(U;F)$ has a unique holomorphic extension $g \in \mathcal{H}(V;F)$.

REMARK 12.2: This proposition states that when E is a complex normed space of dimension greater than or equal to 2 and F is a Banach space, it is not possible for a holomorphic mapping to have an isolated non-removable singularity.

PROOF OF PROPOSITION 12.1: Choose $e \in E$ with $\|e\| = 1$. By the Hahn-Banach theorem there exists $\psi \in E'$ such that $\|\psi\| = 1$ and $\psi(e) = 1$. The kernel S of ψ is a closed hyperplane in E, and so every $x \in E$ can be written uniquely in the form $x = \lambda_x e + s_x$, where $\lambda_x \in \mathbb{C}$ and $s_x \in S$

$(\lambda_x = \psi(x))$.

Let $\rho > 0$ be such that $B_\rho(\xi) \subset V$, and let $\varepsilon = \rho/3$. For $x \in B_\varepsilon(\xi)$ let

$$g(x) = \frac{1}{2\pi i}\int_{|\lambda-\psi(\xi)|=\varepsilon} \frac{f(\lambda e+s_x)}{\lambda-\psi(x)}\, d\lambda$$

$$= \frac{1}{2\pi i}\int_{|\lambda-\psi(\xi)|=\varepsilon} \frac{f[x+(\lambda-\psi(x))e]}{\lambda-\psi(x)}\, d\lambda.$$

To see that the first integral is defined, let $|\lambda-\psi(x)| = \varepsilon$, and $x \in B_\varepsilon(\xi)$. Since $\xi = \psi(\xi)e + s_\xi$,

$$\|\lambda e+s_x-\xi\| = \|\lambda e+s_x-[\psi(\xi)e+s_\xi]\|$$

$$= \|[\lambda-\psi(\xi)]e + x-\xi + [\psi(\xi)-\psi(x)]e\|$$

$$\leq \|[\lambda-\psi(\xi)]e\| + \|x-\xi\| + \|[\psi(\xi)-\psi(x)]e\|$$

$$\leq 3\varepsilon = \rho,$$

and so $\lambda e + s_x \in B_\rho(\xi) \subset V$. Suppose that $\lambda e + s_x = \xi$; then $\lambda = \psi(\lambda e+s_x) = \psi(\xi)$, which is impossible since $|\lambda-\psi(\xi)| = \varepsilon$. Therefore $\lambda e+s_x \in B_\rho(\xi)-\{\xi\} \subset V-\{\xi\} = U$. Finally, $\lambda-\psi(x) = 0$ is impossible, since $\lambda = \psi(x)$ implies $\varepsilon = |\lambda-\psi(\xi)| = |\psi(x)-\psi(\xi)| \leq \|\psi\|\|x-\xi\| = \|x-\xi\| < \varepsilon$. And since F is complete, it follows that $g(x) \in F$ for every $x \in B_\varepsilon(\xi)$.

We assert, without proof, that $g \in \mathcal{H}(B_\varepsilon(\xi);F)$.

Now let $x \in B_\varepsilon(\xi)$ with $s_x \neq s_\xi$. We claim that for every $\lambda \in \mathbb{C}$ with $|\lambda-\psi(\xi)| \leq \varepsilon$, we have $\lambda e+s_x \in B_\rho(\xi)-\{\xi\}\subset$ $\subset U$. The proof that $\lambda e+s_x \in B_\rho(\xi)$ is done in the same way as the proof that g is well defined. Suppose then that

$\lambda e+s_x = \xi$. Then, since $\xi = \psi(\xi)e + s_\xi$, and $\mathbb{C}e + S$ is a direct sum, $s_x = s_\xi$, which is a contradiction. Hence if $\lambda \in \mathbb{C}$, and $|\lambda-\psi(\xi)| \le \varepsilon$, then $\lambda e+s_x \in B_\rho(\xi)-\{\xi\}$.

Now the hypothesis that E has dimension greater than or equal to 2 implies that the set

$$D = \{x \in B_\varepsilon(\xi) : s_x \neq s_\xi\}$$

is not empty. For if $D = \emptyset$ we have $s_x = s_\xi$ for every $x \in B_\varepsilon(\xi)$, and so

$$x = \psi(x)e + s_x = \psi(x)e + s_\xi = \psi(x)e + \xi - \psi(\xi)e$$

$$= \xi + \psi(x-\xi)e \in \xi + \mathbb{C}e$$

for every $x \in B_\varepsilon(\xi)$. Thus the linear variety $\xi + \mathbb{C}e$ has non-empty interior, which implies that $\xi + \mathbb{C}e = E = \mathbb{C}e$, that is, E has dimension 1, contrary to our hypothesis.

We claim that D is dense in $B_\varepsilon(\xi) - \{\xi\}$. Let W be a non-empty open subset of $B_\varepsilon(\xi) - \{\xi\}$, and suppose that $D \cap W = \emptyset$. Then $s_y = s_\xi$ for every $y \in W$, which implies that $y-\xi = [\psi(y)e+s_y] - [\psi(\xi)e+s_\xi] = \psi(y-\xi)e \in \mathbb{C}e$. Therefore $W \subset \xi + \mathbb{C}e$, which is impossible, since E has dimension greater than or equal to 2.

Now given $x \in D$, consider the mapping

$$h: \lambda \in \bar{B}_\varepsilon(\psi(\xi)) \mapsto f[x+(\lambda-\psi(x))e] \in F.$$

It is easy to see that $h \in \mathcal{H}(B_\varepsilon(\psi(\xi));F) \cap C(\bar{B}_\varepsilon(\psi(\xi));F)$. For every $x \in D$ we have

$$|\psi(x)-\psi(\xi)| \le \|x-\xi\| < \varepsilon,$$

and hence $\psi(x) \in B_\varepsilon(\psi(\xi))$. Therefore, applying the Cauchy integral formula to h,

$$h(\psi(x)) = \frac{1}{2\pi i} \int_{|\lambda-\psi(\xi)|=\varepsilon} \frac{h(\lambda)}{\lambda-\psi(x)} \, d\lambda$$

for every $x \in D$. Hence

$$f(x) = \frac{1}{2\pi i} \int_{|\lambda-\psi(\xi)|=\varepsilon} \frac{f[x+(\lambda-\psi(x))e]}{\lambda-\psi(x)} \, d\lambda$$

for every $x \in D$. It follows from the definition of g that $f(x) = g(x)$ for every $x \in D$. Now since f and g are holomorphic, and hence continuous, in $B_\varepsilon(\xi) - \{\xi\}$, and they coincide in the dense subset D, f and g must coincide throughout $B_\varepsilon(\xi) - \{\xi\}$. Hence, if the mapping $g' : V \to F$ is defined by setting $g' = g$ in $B_\varepsilon(\xi)$ and $g' = f$ in $U - \bar{B}_{\varepsilon_1}(\xi)$, with $0 < \varepsilon_1 < \varepsilon$, then $g' \in \mathcal{H}(V;F)$ and $g'|_U = f$. Q.E.D.

REMARK 12.3: If the mapping f of Proposition 12.1 belongs to $\mathcal{H}_b(U;F)$, then $g' \in \mathcal{H}_b(V;F)$.

DEFINITION 12.1 Let U and V be non-empty connected open subsets of E such that $U \subset V$. V is said to be a $\mathcal{H}_b$-holomorphic extension of U if every $f \in \mathcal{H}_b(U)$ possesses a (necessarily unique) extension $g \in \mathcal{H}_b(V)$. V is said to be a proper $\mathcal{H}_b$-holomorphic extension of U if V is a $\mathcal{H}_b$-holomorphic extension of U and $V \supsetneq U$.

We remark that the definition of a $\mathcal{H}_b$-holomorphic extension can be rephrased as follows: the mapping

$$g \in \mathcal{H}_b(V) \mapsto g/U \in \mathcal{H}_b(U)$$

is a surjective isomorphism between the algebras $\mathcal{H}_b(V)$ and $\mathcal{H}_b(U)$.

EXAMPLE 12.1 If E has dimension greater than or equal to 2, V is a connected open subset of E, $\xi \in V$, and $U = V-\{\xi\}$, then by Proposition 12.1 and Remark 12.3, V is a proper $\mathcal{H}_b$-holomorphic extension of U.

REMARK 12.4: In the case where the dimension of E is 1, the definition of a proper $\mathcal{H}_b$-holomorphic extension is of no interest, since the situation described cannot arise. For if U and V are non-empty connected open subsets of E with $U \subsetneq V$, and if $\xi \in V \cap \partial U$, then the function

$$f\colon x \in U \mapsto f(x) = \frac{1}{x-\xi} \in \mathbb{C}$$

belongs to $\mathcal{H}_b(U)$, and has no holomorphic extension to V. Thus V is not a proper $\mathcal{H}_b$-holomorphic extension of U.

Next, we introduce some terminology which will help to simplify the definition of a domain of holomorphy.

DEFINTION 12.2 Let U, V and W be non-empty connected open subsets of E such that $W \subset U \cap V$. V is said to be a $\mathcal{H}_b$-holomorphic prolongation of U by W if for every $f \in \mathcal{H}_b(U)$ there exists $g \in \mathcal{H}_b(V)$ such that $f|_W = g|_W$. V is said to be a proper $\mathcal{H}_b$-holomorphic prolongation of U by W if V is a $\mathcal{H}_b$-holomorphic prolongation of U by W, and $V \not\subset U$.

We say that a non-empty connected open subset U of E has a proper $\mathcal{H}_b$-holomorphic prolongation if there exist non-empty connected open subsets, V and W, of E, such that V is a proper $\mathcal{H}_b$-holomorphic prolongation of U by W.

REMARK 12.5: a) If V is a $\mathcal{H}_b$-holomorphic prolongation of U by W, then for every $f \in \mathcal{H}_b(U)$ there exists $g \in \mathcal{H}_b(V)$ such that g coincides with f in the connected component W_o of $U \cap V$ which contains W.

b) We point out that Definition 12.1 is a special case of Definition 12.2, in the following sense: if U and V are non-empty connected open subsets of E, $U \subset V$ and V is a (proper) $\mathcal{H}_b$-holomorphic extension of U, then V is a (proper) $\mathcal{H}_b$-holomorphic prolongation of U by U.

DEFINITION 12.3 (Domains of $\mathcal{H}_b$-holomorphy). A non-empty connected open subset U of E is a domain of $\mathcal{H}_b$-holomorphy if U has no proper $\mathcal{H}_b$-holomorphic prolongation. Explicitly, this means that there does not exist a pair of connected open sets V and W such that

1) $\emptyset \neq W \subset U \cap V$.

2) For every $f \in \mathcal{H}_b(U)$ there exists $g \in \mathcal{H}_b(V)$ such that $g|_W = f|_W$.

EXAMPLE 12.2 Let V be a non-empty connected open subset of E, $\xi \in V$, $U = V-\{\xi\}$. If $\dim E \geq 2$ then V is a proper $\mathcal{H}_b$-holomorphic extension of U (Proposition 12.1), and hence V is a proper $\mathcal{H}_b$-holomorphic prolongation of U by U. Therefore U is not a domain of $\mathcal{H}_b$-holomorphy.

EXAMPLE 12.3 Every non-empty connected open subset U of $\mathbb{C}$

is a domain of holomorphy that is, a domain of $\mathcal{H}_b$-holomorphy. For suppose that U has a proper $\mathcal{H}_b$-prolongation. Then there exist non-empty connected open sets, V and W, such that $W \subset U \cap V$, $V \not\subset U$, and for every $f \in \mathcal{H}(U)$ there exists $g \in \mathcal{H}(V)$ such that $f|_W = g|_W$. Then, if W_o is the connected component of $U \cap V$ containing W, we have $V \cap \partial U \cap \partial W_o \neq \neq \phi$ (see the proof of Proposition 12.2). If we choose $\xi \in V \cap \partial U \cap \partial W_o$, then the function $f: x \in U \mapsto \frac{1}{x-\xi} \in \mathbb{C}$ belongs to $\mathcal{H}(U)$, and there is no $g \in \mathcal{H}(V)$ such that $f|_W = g|_W$, since if such a g existed, we would have $f = g$ in W_o. But then since $\xi \in \bar{W}_o$ and g is continuous at ξ,

$$|g(\xi)| = \lim_{\substack{x \to \xi \\ x \in W_o}} |g(x)| = \lim_{\substack{x \to \xi \\ x \in W_o}} |f(x)| = \infty,$$

which is impossible, as $g(\xi) \in \mathbb{C}$. Therefore U is a domain of holomorphy.

This example is the inspiration for the following proposition.

PROPOSITION 12.2 Suppose that U is a non-empty connected open subset of E, such that for every $\xi \in \partial U$ there exists $f \in \mathcal{H}_b(U)$ for which $\lim_{\substack{x \to \xi \\ x \in U}} |f(x)| = \infty$.

Then U is a domain of $\mathcal{H}_b$-holomorphy.

PROOF: Suppose that U is not a domain of $\mathcal{H}_b$-holomorphy. Then U has a proper $\mathcal{H}_b$-holomorphic prolongation V by a set W. Let W_o be the connected component of $U \cap V$ containing W, let $a \in W$ and $b \in V \setminus U$. Let γ be a path in

V from a to b; thus γ is a continuous mapping from $[0,1]$ into V, with $\gamma(0) = a$ and $\gamma(1) = b$. Let (γ) denote the image of γ. Now $(\gamma) \cap \partial W_o$ is a closed subset of V, and it is non-empty since (γ) is connected. Therefore the set of points in $[0,1]$ corresponding to points in $(\gamma) \cap \partial W_o$ is closed. It follows that this set has a least element; that is, there exists $t_o \in [0,1]$ such that $\gamma(t_o) \in \partial W_o$, and $\gamma(t) \notin \partial W_o$ for all $t \in [0,t_o)$. Let $\gamma(t_o) = \xi$.

We claim that $\xi \in \partial U$. If this is not so there are two possibilities:

1) $\xi \in U$. Then there is a neighbourhood $V(\xi)$ of ξ contained in $U \cap V$. By the choice of ξ, this implies that $V(\xi)$ contains points in W_o and points outside W_o; that is, $W_o \subsetneqq W_o \cup V(\xi)$. But $V(\xi)$ may be chosen to be connected, and then $W_o \cup V(\xi)$ is a connected set, contained in $U \cap V$, and containing W_o as a proper subset. This is impossible since W_o is a connected component of $U \cap V$.

2) $\xi \notin U$. Then $\xi \notin U \cup \partial U = \bar{U}$, implying that $\xi \notin \bar{W}_o$, which is absurd.

Therefore $\xi \in \partial U \cap (\gamma) \cap \partial W_o$. By hypothesis, there then exists $f \in \mathcal{H}_b(U)$ such that $\lim\limits_{\substack{x \to \xi \\ x \in U}} |f(x)| = \infty$.

If V is a proper $\mathcal{H}_b$-holomorphic prolongation of U by W, there exists $g \in \mathcal{H}_b(V)$ such that $g|_W = f|_W$. Then $g|_{W_o} = f|_{W_o}$, and since $\xi \in V \cap \partial W_o$,

$$|g(\xi)| = \lim_{\substack{x\to\xi \\ x\in W_o}} |g(x)| = \lim_{\substack{x\to\xi \\ x\in W_o}} |f(x)| = \infty,$$

which is absurd. Q.E.D.

DEFINITION 12.4 A subset X of E is said to be subconvex if every $x \in E\backslash X$ is contained in a complex affine variety S of codimension 1, with $S \cap X = \emptyset$.

REMARK 12.6: a) We note that every complex affine variety S of codimension 1 is of the form $S = \eta + T$, where $\eta \in E$ and T is a complex vector subspace of E of codimension 1.

b) If U is a non-empty open subconvex subset of E and $\xi \in E\backslash U$, the complex affine variety S of codimension 1 which passes through ξ and does not intersect U is necessarily closed, since it would otherwise be dense in E, and as U is open, this is impossible.

PROPOSITION 12.3 Every open convex subset U of E is subconvex.

PROOF: If $U = \emptyset$ this is clear. If $U \neq \emptyset$ and $\xi \in E\backslash U$, then by the Hahn-Banach theorem applied to the space E, considered as a real vector spaces there exists a real vector subspace T of E, of real codimension 1, such that $(\xi+T) \cap U = \emptyset$. Let $S = \xi + (T\cap iT)$. Then S is a complex affine variety of complex codimension 1, $\xi \in S$, and $S \cap U = \emptyset$. Q.E.D.

REMARK 12.7 Every subset of $\mathbb{C}$, convex or not, is subconvex. For if $X \subset \mathbb{C}$ and $\xi \notin X$, we may take $S = \xi + \{0\}$, since $\{0\}$ is a subspace of $\mathbb{C}$ of codimension 1.

PROPOSITION 12.4 Every subconvex connected open subset U of E is a domain of $\mathcal{H}_b$-holomorphy.

PROOF: Let $\xi \in \partial U$. Since U is open, $\xi \notin U$, and so, U being subconvex, there exists a complex affine variety $S = \eta + T$ of codimension 1, such that $\xi \in S$ and $S \cap U = \emptyset$. By Remark 12.6 (b) S, and hence T, is closed. Therefore there exists $\varphi \in E'$, the topological dual of E, for which $T = \{x \in E : \varphi(x) = 0\}$. Since $\xi \in S$, $\xi - \eta \in T$, and so $\varphi(\xi) = \varphi(\eta)$. Thus $S = \{x \in E : x-\eta \in T\} = \{x \in E : \varphi(x) = \varphi(\xi) = \lambda\}$; and if $x \in U$ then $x \notin S$, which implies that $\varphi(x) \neq \lambda$. It follows that the mapping

$$f: x \in U \mapsto \frac{1}{\varphi(x)-\lambda} \in \mathbb{C}$$

belongs to $\mathcal{H}_b(U)$, and $\lim_{\substack{x \to \xi \\ x \in U}} |f(x)| = \infty$. Therefore, by Proposition 12.2, U is a domain of $\mathcal{H}_b$-holomorphy. Q.E.D.

By Proposition 12.3, we deduce:

COROLLARY 12.1 Every convex connected open subset of E is a domain of $\mathcal{H}_b$-holomorphy.

CHAPTER 13

THE CARTAN-THULLEN THEOREM FOR DOMAINS OF $\mathcal{H}_b$-HOLOMORPHY

DEFINITION 13.1 Let U be a non-empty open subset of E, and let $X \subset U$. The $\mathcal{H}_b$-hull of X is the subset of U defined by:

$$\hat{X}_{\mathcal{H}_b(U)} = \{t \in U : |f(t)| \leq \sup_{x \in X} |f(x)| \text{ for every } f \in \mathcal{H}_b(U)\}.$$

PROPOSITION 13.1

(1) $X \subset \hat{X}_{\mathcal{H}_b(U)}$.

(2) $\hat{X}_{\mathcal{H}_b(U)}$ is closed in U for every $X \subset U$.

(3) $\hat{\emptyset}_{\mathcal{H}_b(U)} = \emptyset$, $\hat{U}_{\mathcal{H}_b(U)} = U$.

(4) If $X \subset Y \subset U$, then $\hat{X}_{\mathcal{H}_b(U)} \subset \hat{Y}_{\mathcal{H}_b(U)}$.

(5) If U, V are non-empty open subsets of E with $U \subset V$, then $\hat{X}_{\mathcal{H}_b(U)} \subset \hat{X}_{\mathcal{H}_b(V)}$ for every $X \subset U$.

(6) $\sup_{t \in \hat{X}_{\mathcal{H}_b(U)}} |f(t)| = \sup_{x \in X} |f(x)|$ for every $f \in \mathcal{H}_b(U)$.

(7) $\widehat{\left[\hat{X}_{\mathcal{H}_b(U)}\right]}_{\mathcal{H}_b(U)} = \hat{X}_{\mathcal{H}_b(U)}$ for every $X \subset U$.

(8) If $\bar{X}$ is the closure of X in U then $\widehat{(\bar{X})}_{\mathcal{H}_b(U)} = \hat{X}_{\mathcal{H}_b(U)}$.

PROOF: We shall prove (2) and (6) only.

(2) $$\hat{X}_{\mathcal{H}_b(U)} = \bigcap_{f \in \mathcal{H}_b(U)} \{t \in U : |f(t)| \leq \sup_{x \in X} |f(x)|\};$$

thus $\hat{X}_{\mathcal{H}_b(U)}$ is the intersection of a family of closed subsets of U.

(6) $\hat{X}_{\mathcal{H}_b(U)}$ can be described as the largest of the subsets T of U with the property:

$$\sup_{t \in T} |f(t)| = \sup_{x \in X} |f(x)| \quad \text{for every} \quad f \in \mathcal{H}_b(U).$$

For if $t \in \hat{X}_{\mathcal{H}_b(U)}$, then for every $f \in \mathcal{H}_b(U)$ we have

$$|f(t)| \leq \sup_{x \in X} |f(x)|,$$

which implies that $\sup_{t \in \hat{X}_{\mathcal{H}_b(U)}} |f(t)| \leq \sup_{x \in X} |f(x)|$.

But as $X \subset \hat{X}_{\mathcal{H}_b(U)}$, the reverse inequality also holds.

On the other hand, if $T \subset U$ is such that

$$\sup_{t \in T} |f(t)| = \sup_{x \in X} |f(x)| \quad \text{for every} \quad f \in \mathcal{H}_b(U),$$

then for every $t \in T$, $|f(t)| \leq \sup_{x \in X} |f(x)|$ for every $f \in \mathcal{H}_b(U)$, and therefore $T \subset \hat{X}_{\mathcal{H}_b(U)}$. Q.E.D.

REMARK 13.1: Let $f \in \mathcal{H}_b(U)$, and suppose that $|f(x)| \leq C$ for every $x \in X$. By Definition 13.1 we have that $|f(x)| \leq C$ for every $x \in \hat{X}_{\mathcal{H}_b(U)}$. Thus, an inequality of the form $|f| \leq C$ which is valid in X, is valid in $\hat{X}_{\mathcal{H}_b(U)}$. This is the fundamental property of the $\mathcal{H}_b(U)$-hull of a subset X of U.

PROPOSITION 13.2 If X is a subset of E, then $\hat{X}_{\mathcal{H}_b(E)}$ is contained in the closed convex balanced hull $\hat{X}$, of X in E.

PROOF: We shall employ the following formulation of the Hahn-Banach theorem:

(*) Let E be a real normed space, $X \subset E$ and $t \in X$. Then t belongs to the closed convex balanced hull of X in E if and only if $\varphi(t) \le \sup_{x \in X} \varphi(x)$ for every continuous linear form φ on E.

Let $E_{\mathbb{R}}$ denote the real space associated with E, and let $\varphi \in (E_{\mathbb{R}})'$, the topological dual of $E_{\mathbb{R}}$. Define $\Phi: E \mapsto \mathbb{C}$ by $\Phi(x) = \varphi(x) - i\varphi(ix)$, $x \in E$. Then $\Phi \in E'$, that is, Φ is a continuous complex-linear form on E. Consider the mapping $e^{\Phi}: x \in E \mapsto e^{\Phi(x)} \in \mathbb{C}$. The mapping e^{Φ} is an entire function on E, being the composition of the continuous linear function Φ and the entire function $e^z: \mathbb{C} \to \mathbb{C}$. Also, e^{Φ} is of bounded type, since if $A \subset E$ is bounded, and $M = \sup_{x \in A} \|x\|$, then

$$|e^{\Phi(x)}| \le e^{M\|\varphi\|}$$

for every $x \in A$. Therefore, if $t \in \hat{X}_{\mathcal{H}_b(E)}$, $|e^{\Phi(t)}| \le$ $\le \sup_{x \in X} |e^{\Phi(x)}|$, which implies that $e^{\varphi(t)} \le \sup_{x \in X} e^{\varphi(x)}$. Hence $\varphi(t) \le \sup_{x \in X} \varphi(x)$ for every $\varphi \in (E_{\mathbb{R}})'$, and it follows from (*) that t belongs to the closed convex balanced hull, $\hat{X}$, of X. Q.E.D.

COROLLARY 13.1 If X is bounded (respectively, precompact), then $\hat{X}_{\mathcal{H}_b(E)}$ is also bounded (respectively, precompact). More generally, if X is bounded (respectively, precompact), and X is a subset of the non-empty open subset U of E, then $\hat{X}_{\mathcal{H}_b(U)}$ is also bounded (respectively, precompact).

PROOF: These statements all follow from the fact that $X \subset \hat{X}_{\mathcal{H}_b(U)} \subset \hat{X}_{\mathcal{H}_b(E)} \subset \hat{X}$, since $\hat{X}$ is bounded (respectively, precompact) when X is bounded (respectively, precompact).

Q.E.D.

REMARK 13.2: The $\mathcal{H}_b(U)$-hull of a subset X of U is closed in U, but is not necessarily closed in E, even if X is compact. The following example illustrates this phenomenon.

EXAMPLE 13.1 Let $E = \mathbb{C}^n$, $n \geq 2$, $\xi \in \mathbb{C}^n$ and $U = \mathbb{C}^n - \{\xi\}$. Let

$$X = \{z \in \mathbb{C}^n : \|z-\xi\| = [\sum_{i=1}^{n} |z_i - \xi_i|^2]^{1/2} = 1\}.$$

Then X is compact. If $f \in \mathcal{H}(U)$ then, by Proposition 12.1, f has a holomorphic extension to $f \in \mathcal{H}(E)$. By the maximum norm theorem (Proposition 10.2) we then have $|f(z)| \leq$ $\leq \sup_{x \in X} |f(x)|$ for every $z \in \mathbb{C}^n$, $\|z-\xi\| \leq 1$.
Therefore every $z \in \mathbb{C}^n$ with $0 < \|z-\xi\| \leq 1$ belongs to $\hat{X}_{\mathcal{H}_b(U)}$, but since $\xi \notin U$, $\xi \notin \hat{X}_{\mathcal{H}_b(U)}$. Thus $\hat{X}_{\mathcal{H}_b(U)}$ is not closed in $E = \mathbb{C}^n$.

DEFINITION 13.2 A non-empty open subset U of E is said to be $\mathcal{H}_b$-holomorphically convex if for every subset X of U which is U-bounded, its $\mathcal{H}_b(U)$-hull, $\hat{X}_{\mathcal{H}_b(U)}$, is also U-bounded.

We recall that to say $X \subset U$ is U-bounded means that $\text{dist}(X, \partial U) > 0$, and X is bounded in E.

LEMMA 13.1 If $f \in \mathcal{H}_b(U;F)$, then for every $m \in \mathbb{N}$, $\hat{d}^m f \in \mathcal{H}_b(U; \mathcal{P}(^m E;F))$ and $d^m f \in \mathcal{H}_b(U; \mathcal{L}_s(^m E;F))$.

PROOF: If $f \in \mathcal{H}_b(U;F)$ then, by Proposition 5.4, $\hat{d}^m f \in \mathcal{H}(U;\mathcal{P}(^mE;F))$ and $d^m f \in \mathcal{H}(U;\mathcal{L}_s(^mE;F))$ for every $m \in \mathbb{N}$. Thus we have only to prove that $\hat{d}^m f$ and $d^m f$ are bounded on the U-bounded subsets of U.

Let $m \in \mathbb{N}$, and let X be a U-bounded subset of U, so that $\operatorname{dist}(X,\partial U) = d > 0$. Let $\rho = d/2$. We claim that the set $\bar{B}_\rho(X) = \bigcup_{x\in X} \bar{B}_\rho(x)$ is U-bounded. It is clear that $\bar{B}_\rho(X)$ is bounded, and that $\bar{B}_\rho(X) \subset U$. Let $z \in \bar{B}_\rho(X)$, and $u \in \partial U$. We have

$$\operatorname{dist}(u,X) \geq d > \rho \geq \operatorname{dist}(z,X).$$

Therefore $\operatorname{dist}(z,U) \geq |\operatorname{dist}(u,X)-\operatorname{dist}(z,X)| \geq d-\rho = \rho$, and so $\operatorname{dist}(\bar{B}_\rho(X),\partial U) \geq \rho$.

Since $\bar{B}_\rho(X)$ is U-bounded, and $f \in \mathcal{H}_b(U;F)$, $\sup_{y\in\bar{B}_\rho(X)} \|f(y)\| < \infty$. For every $x \in X$, $\bar{B}_\rho(x) \subset \bar{B}_\rho(X) \subset U$, and so by the Cauchy inequalities,

$$\left\|\frac{1}{m!}\hat{d}^m f(x)\right\| \leq \frac{1}{\rho^m} \sup_{y\in\bar{B}_\rho(x)} \|f(y)\|$$

for every $x \in X$, $m \in \mathbb{N}$. Therefore

$$\sup_{x\in X} \left\|\frac{1}{m!}\hat{d}^m f(x)\right\| \leq \frac{1}{\rho^m} \sup_{x\in X} \{ \sup_{y\in\bar{B}_\rho(x)} \|f(y)\| \}$$

$$= \frac{1}{\rho^m} \sup_{y\in\bar{B}_\rho(X)} \|f(y)\| < \infty,$$

and so

1) $$\sup_{x\in X} \|\hat{d}^m f(x)\| \leq \frac{m!}{\rho^m} \sup_{y\in\bar{B}_\rho(X)} \|f(y)\| < \infty$$

for every $m \in \mathbb{N}$. Also, by Proposition 1.3, we have

$$2) \quad \sup_{x\in X} \|d^m f(x)\| \le \frac{m^m}{\rho^m} \sup_{y\in \bar{B}_\rho(X)} \|f(y)\| < \infty$$

for every $m \in \mathbb{N}$. Q.E.D.

LEMMA 13.2 Let X be a U-bounded subset of U, and let $r = \text{dist}(X,\partial U)$ (thus $r > 0$). Then for every $f \in \mathcal{H}_b(U)$ and for every $t \in \hat{X}_{\mathcal{H}_b(U)}$ the radius of convergence of the Taylor series of f at t is greater than or equal to r.

PROOF: Let $\theta \in R$, $0 < \theta < 1$; then, from the proof of Lemma 13.1, $\bar{B}_{\theta r}(X)$ is U-bounded, and so if $f \in \mathcal{H}_b(U)$,

$$M(\theta) = \sup\{|f(x)| : x \in \bar{B}_{\theta r}(X)\} < \infty.$$

Applying the Cauchy inequalities, we have

$$\|\frac{1}{m!} \hat{d}^m f(x)\| \le \frac{M(\theta)}{(\theta r)^m} \quad \text{for every} \quad x \in X, \quad m \in \mathbb{N}.$$

Thus

$$(*) \quad |\frac{1}{m!}\hat{d}^m f(x)(y)| \le \frac{M(\theta)}{(\theta r)^m} \|y\|^m \text{ for every } x \in X,\ y \in E,\ m \in \mathbb{N}.$$

By Lemma 13.1, $\hat{d}^m f \in \mathcal{H}_b(U;\mathcal{P}(^m E))$, and this implies that, for every $y \in E$, the function

$$\hat{d}^m f(\cdot)(y):\ x \in U \mapsto \hat{d}^m f(x)(y) \in \mathbb{C}$$

is of bounded type. For if T is U-bounded, then

$$\sup_{x\in T} |\hat{d}^m f(x)(y)| \le \sup_{x\in T} \|\hat{d}^m f(x)\| \|y\|^m < \infty.$$

Now Remark 13.1 applied to the function $\hat{d}^m f(\cdot)(y) \in$ $\in \mathcal{H}_b(U)$ shows that inequality (*) is valid in $\hat{X}_{\mathcal{H}_b(U)}$. Therefore

$$\left|\frac{1}{m!}\hat{d}^m f(x)(y)\right| \le \frac{M(\theta)}{(\theta r)^m}\|y\|^m$$

for every $x \in \hat{X}_{\mathcal{H}_b(U)}$, $y \in E$ and $m \in \mathbb{N}$. Therefore

$$\left\|\frac{1}{m!}\hat{d}^m f(x)\right\| \le \frac{M(\theta)}{(\theta r)^m} \quad \text{for every } x \in \hat{X}_{\mathcal{H}_b(U)} \text{ and } m \in \mathbb{N}.$$

Now fix a point $t \in \hat{X}_{\mathcal{H}_b(U)}$, and let ρ be the radius of convergence of the Taylor series of f at t. Applying the Cauchy-Hadamard formula to the above,

$$\frac{1}{\rho} = \limsup_{m\to\infty} \left\|\frac{1}{m!}\hat{d}^m f(t)\right\|^{1/m}$$

$$\le \limsup_{m\to\infty} \frac{M(\theta)^{1/m}}{r\theta} = \frac{1}{r\theta}.$$

Therefore $\rho \ge r\theta$ for every $\theta \in \mathbb{R}$, $0 < \theta < 1$. Hence $\rho \ge r$. Q.E.D.

REMARK 13.3: If $P \in \mathcal{P}(^mE;F)$, it is clear that $P/W \in \mathcal{H}_b(W;F)$ for every non-empty open subset W of E. Hence, if $P \in \mathcal{P}(E;F)$, then $P|_W \in \mathcal{H}_b(W;F)$ for every non-empty open set $W \subset E$.

LEMMA 13.3 Let E and F be normed spaces, F complete, $\xi \in E$, and $P_m \in \mathcal{P}(^mE;F)$ for every $m \in \mathbb{N}$. If the series $f(x) = \sum_{m=0}^{\infty} P_m(x-\xi)$ has radius of convergence $r > 0$, then $f \in \mathcal{H}_b(B_r(\xi);F)$.

PROOF: It is easy to see that it is sufficient to consider the case $\xi = 0$. Let $U = B_r(0)$, and for each $m \in \mathbb{N}$, let $f_m: U \to F$ be the mapping defined by $f_m = \sum_{k=0}^{m} P_k/U$, that is $f_m(x) = \sum_{k=0}^{m} P_k(x)$ for $x \in U$. By Remark 13.3, $f_m \in \mathcal{H}_b(U;F)$ for every $m \in \mathbb{N}$. By the definition of the radius of convergence, the sequence $\{f_m\}_{m\in\mathbb{N}}$ converges uniformly to f in $\bar{B}_\rho(0)$ for every $\rho \in [0,r)$, and since every U-bounded set is contained in $\bar{B}_\rho(0)$ for some $\rho \in [0,r)$, it follows that $\{f_m\}_{m\in\mathbb{N}}$ is a Cauchy sequence in $\mathcal{H}_b(U;F)$ with the natural topology. Since F is complete, we have, by Proposition 8.1, that $\mathcal{H}_b(U;F)$ is complete, and hence $\{f_m\}_{m\in\mathbb{N}}$ converges in the natural topology to some $g \in \mathcal{H}_b(U;F)$. In particular, $\{f_m\}_{m\in\mathbb{N}}$ converges pointwise in U to g. Therefore $g = f$, and $f \in \mathcal{H}_b(U;F)$. Q.E.D.

THE CARTAN-THULLEN THEOREM (part I): Let U be a non-empty connected open subset of E. Then the following are equivalent:

a) U is a domain of $\mathcal{H}_b$-holomorphy.

b) For every U-bounded subset X of U,

$$\mathrm{dist}(X,\partial U) = \mathrm{dist}(\hat{X}_{\mathcal{H}_b(U)},\partial U).$$

c) U is $\mathcal{H}_b$-holomorphically convex.

d) For every sequence $\{\xi_n\}_{n\in\mathbb{N}}$ in U which converges to a point $\xi \in \partial U$ there exists $f \in \mathcal{H}_b(U)$ which is unbounded on $\{\xi_n\}_{n\in\mathbb{N}}$, that is, $\sup_{n\in\mathbb{N}} |f(\xi_n)| = \infty$.

PROOF: a) $\Rightarrow$ b). Since $X \subset \hat{X}_{\mathcal{H}_b(U)}$, it is clear that $\mathrm{dist}(X,\partial U) \geq \mathrm{dist}(\hat{X}_{\mathcal{H}_b(U)},U)$.

Let $\mathrm{dist}(X,\partial U) = r > 0$. If $t \in \hat{X}_{\mathcal{H}_b(U)}$, the set $V = B_r(t)$ is a non-empty connected open subset of E. Denoting by W the connected component of $U \cap V$ which contains t, U, V and W are all non-empty connected open sets, and $U \subset U \cap V$. Now if $f \in \mathcal{H}_b(U)$, there exists a real number $\sigma > 0$, depending on f, such that $\bar{B}_\sigma(t) \subset U$, and

$$f(x) = \sum_{m=0}^{\infty} \frac{1}{m!} \hat{d}^m f(t)(x-t)$$

uniformly in $\bar{B}_\sigma(t)$. If ρ is the radius of convergence of this Taylor series, then by Lemma 13.2, $\rho \geq r$, and by Lemma 13.3, the function

$$g_o\colon x \in B_\rho(t) \mapsto g_o(x) = \sum_{m=0}^{\infty} \frac{1}{m!} \hat{d}^m f(t)(x-t)$$

is an element of $\mathcal{H}_b(B_\rho(t))$. Since $\rho \geq r$ we may restrict g to V to obtain a function $g = g_o\big|_V \in \mathcal{H}_b(V)$. From the definition of g_o we have $f(x) = g(x)$ for every $x \in B_\sigma(t)$, and hence $f\big|_W = g\big|_W$.

Thus, we have shown that for every $f \in \mathcal{H}_b(U)$ there exists $g \in \mathcal{H}_b(V)$ such that $f\big|_W = g\big|_W$. But if $V \not\subset U$, this means that V is a $\mathcal{H}_b$-holomorphic prolongation of U by W, which is impossible since U is a domain of $\mathcal{H}_b$-holomorphy. Therefore V must be a subset of U, in other words, $B_r(t) \subset U$ for every $t \in \hat{X}_{\mathcal{H}_b(U)}$, and so

$$\operatorname{dist}(\hat{X}_{\mathcal{H}_b(U)}, \partial U) \geq r = \operatorname{dist}(X, \partial U).$$

b) $\Rightarrow$ c). Let X be a U-bounded set. Then, in particular, X is bounded, and so by the Corollary 13.1, $\hat{X}_{\mathcal{H}_b(U)}$ is bounded. Now by b)

$$\operatorname{dist}(X, \partial U) = \operatorname{dist}(\hat{X}_{\mathcal{H}_b(U)}, \partial U),$$

and so since $\operatorname{dist}(X, \partial U) > 0$, we have also $\operatorname{dist}(\hat{X}_{\mathcal{H}_b(U)}, \partial U) > 0$.

Therefore $\hat{X}_{\mathcal{H}_b(U)}$ is U-bounded for every U-bounded set X. Hence U is holomorphically convex.

c) ⇒ d). Suppose d) is false. Then there is a sequence $\{\xi_n\}_{n\in N}$ of points in U which converges to a point $\xi \in \partial U$, such that every $f \in \mathcal{H}_b(U)$ is bounde on the set $\{\xi_n : n \in N\}$. It follows that the mapping

$$p: f \in \mathcal{H}_b(U) \mapsto p(f) = \sup_{n\in\mathbb{N}} |f(\xi_n)| \in \mathbb{R}$$

is a seminorm on $\mathcal{H}_b(U)$. Since the linear form $f \in \mathcal{H}_b(U) \mapsto$ $\mapsto f(x) \in \mathbb{C}$ is continuous for every $x \in U$, and p is the supremum of a sequence of the modulus of those linear forms p is lower semicontinuous. By Proposition 8.1 we have that $\mathcal{H}_b(U)$ with the natural topology is a Fréchet space. In particular, it is barreled, which implies that every lower semicontinuous seminorm is continuous. Therefore p is a continuous seminorm on $\mathcal{H}_b(U)$ for the natural topology. It follows that there exist a U-bounded set X and a real number $C > 0$ such that

$$(*) \qquad p(f) \leq Cp_X(f) = C \sup_{x\in X} |f(x)|$$

for every $f \in \mathcal{H}_b(U)$. Since $\mathcal{H}_b(U)$ is an algebra over $\mathbb{C}$ we may replace f by f^m $(m \in \mathbb{N})$ in (*):

$$p(f^m) \leq Cp_X(f^m) \quad \text{for every} \quad f \in \mathcal{H}_b(U), \quad m \in \mathbb{N}.$$

Thus

$$p(f^m) = \sup_{n\in\mathbb{N}} |f(\xi_n)|^m = [\sup_{n\in\mathbb{N}} |f(\xi_n)|]^m = [p(f)]^m$$
$$\leq Cp_X(f^m) = C[p_X(f)]^m$$

for every $f \in \mathcal{H}_b(U)$, $m \in \mathbb{N}$. Therefore

$$(**) \qquad p(f) \leq C^{1/m} \, p_X(f)$$

for every $f \in \mathcal{H}_b(U)$, $m \in \mathbb{N}$. Letting m tend to ∞, we obtain

$$p(f) \leq p_X(f)$$

for every $f \in \mathcal{H}_b(U)$. It follows from the definition of p that

$$|f(\xi_n)| \leq \sup_{x \in X} |f(x)|$$

for every $f \in \mathcal{H}_b(U)$, $n \in \mathbb{N}$, and so, by the definition of the $\mathcal{H}_b(U)$-hull, $\xi_n \in \hat{X}_{\mathcal{H}_b(U)}$ for every $n \in \mathbb{N}$. But by hypothesis, $\xi_n \to \xi \in \partial U$ as $n \to \infty$, which implies that $\text{dist}(\hat{X}_{\mathcal{H}_b(U)}, \partial U) = 0$. Therefore $\hat{X}_{\mathcal{H}_b(U)}$ is not U-bounded, which contradicts c).

d) $\Rightarrow$ a). Suppose that a) is false. Then there exist non-empty connected open sets V and W, with $W \subset U \cap V$, $V \not\subset U$, such that for every $f \in \mathcal{H}_b(U)$ there exists $g \in \mathcal{H}_b(V)$ for which $f|_W = g|_W$. Choose $a \in W$ and $b \in V \setminus U$, and let $\Gamma: [0,1] \to V$ be a path with $\Gamma(0) = a$, $\Gamma(1) = b$. Let ξ be the first point on ∂W_o which lies in the image of Γ (see the proof of Proposition 12.3). Then $\xi \in V \cap \partial U \cap \partial W_o$. Now let $\{\xi_n\}_{n \in \mathbb{N}}$ be a sequence of points belonging to the image of Γ, and strictly preceding ξ, such that $\xi_n \to \xi$ as $n \to \infty$. This can be accomplished by choosing a sequence $\{t_n\}_{n \in \mathbb{N}} \subset [0,t)$, where $\Gamma(t) = \xi$, such that $t_n \to T$ as $n \to \infty$, and taking $\xi_n = \Gamma(t_n)$.

We then have $\xi_n \in W_o \subset U$ for every $n \in \mathbb{N}$ and so by

d) there exists $f \in \mathcal{H}_b(U)$ such that $\sup_{n\in\mathbb{N}} |f(\xi_n)| = \infty$. However, by our initial assumption, there exists $g \in \mathcal{H}_b(V)$ such that $f|_W = g|_W$. This implies that $f|_{W_o} = g|_{W_o}$, and since $\xi \in \partial W_o$, we have

$$g(\xi) = \lim_{n\to\infty} g(\xi_n) = \lim_{n\to\infty} f(\xi_n),$$

which is absurd, since $|g(\xi)|$ is finite, and $\lim_{n\to\infty} f(\xi_n) = \infty$.

Q.E.D.

DEFINITION 13.3 Let U be a non-empty connected open set, $f \in \mathcal{H}_b(U)$, and $\xi \in \partial U$. f is said to be $\mathcal{H}_b$-regular at ξ if there exists a pair of non-empty connected open sets V, W, such that $W \subset U \cap V$, $\xi \in V$ (which implies that $V \not\subset U$), and there exists $g \in \mathcal{H}_b(V)$ such that $g|_W = f|_W$. Conversely, ξ is said to be a $\mathcal{H}_b$-singular point for f if no such pair of sets exist. f is said to be $\mathcal{H}_b$-singular on ∂U if every point of ∂U is a $\mathcal{H}_b$-singular point of f. This means that for all non-empty open subsets V, W of E with $W \subset U \cap V$ and $V \not\subset U$, there is no $g \in \mathcal{H}_b(V)$ for which $g = f$ in W.

$S_b(U)$ will denote the set of all $f \in \mathcal{H}_b(U)$ which are $\mathcal{H}_b$-singular at every point of ∂U. U is said to be a $\mathcal{H}_b$-domain of existence if $S_b(U) \neq \emptyset$.

THE CARTAN-THULLEN THEOREM (part II): Suppose that E is separable, and let U be a non-empty connected open subset of E. Then the following are equivalent:

a) U is a domain of $\mathcal{H}_b$-holomorphy.

b) U is a $\mathcal{H}_b$-domain of existence.

c) The complement $\complement S_b(U)$ of $S_b(U)$ in $\mathcal{H}_b(U)$ is of first category in $\mathcal{H}_b(U)$.

In order to prove this theorem, we need the following propositions:

PROPOSITION 13.3 (Montel's Theorem). If E is separable, and $\{f_n\}_{n\in\mathbb{N}}$ is a sequence in $\mathcal{H}_b(U)$ such that $\sup_{x\in X, n\in\mathbb{N}} |f_n(x)| < \infty$ for every U-bounded set X (in particular, if $\sup_{x\in U, n\in\mathbb{N}} |f_n(x)| < \infty$), then there is a subsequence of $\{f_n\}_{n\in\mathbb{N}}$ which converges uniformly on every compact subset of U to a function $f \in \mathcal{H}_b(U)$.

LEMMA 13.4 (Ascoli). Let M be a separable metric spaces and $\{f_n\}_{n\in\mathbb{N}}$ a sequence of complex functions on M. Suppose that the sequence $\{f_n\}_{n\in\mathbb{N}}$ is equicontinuous, and that $\sup_{n\in\mathbb{N}} |f_n(x)| < \infty$ for every $x \in M$. Then there is a subsequence of $\{f_n\}_{n\in\mathbb{N}}$ which converges uniformly on every compact subset of M to a continuous function on M.

PROOF OF PROPOSITION 13.3: Since E is separable, U is a separable metric space. We claim that the sequence $\{f_n\}_{n\in\mathbb{N}}$ is equicontinuous in U. To see this, let $\xi \in U$ and $r \in \mathbb{R}$, $0 < r < \mathrm{dist}(\xi,\partial U)$. By Corollary 4.2 we have

$$|f_n(x)-f_n(\xi)| \leq \frac{\|x-\xi\|}{r-\|x-\xi\|} \sup_{\|t-\xi\|=r} |f_n(t)|.$$

Since $\{t \in E : \|t-\xi\| = r\}$ is U-bounded,

$$\sup\{\|f_n(t)\| : \|t-\xi\| = r, \ n \in \mathbb{N}\} = C < \infty.$$

Therefore $|f_n(x)-f_n(\xi)| \leq C \dfrac{\|x-\xi\|}{r-\|x-\xi\|}$ for every $n \in \mathbb{N}$, which shows that $\{f_n\}_{n\in\mathbb{N}}$ is equicontinuous. Also, by hypothesis, $\sup\limits_{n\in\mathbb{N}} |f_n(x)| < \infty$ for every $x \in U$, and so by Lemma 13.4 there is a subsequence $\{f_{n_k}\}_{k\in\mathbb{N}}$ which converges uniformly on every compact subset of U to a continuous function f. f is holomorphic, since $\mathcal{H}(U)$ is closed in $C(U)$ for the compact-open topology, and since $\sup\limits_{x\in X, k\in\mathbb{N}} |f_{n_k}(x)| < \infty$ for every U-bounded set X, it follows that $f \in \mathcal{H}_b(U)$.

REMARK 13.4: Montel's Theorem can be rephrased as follows: every subset of $\mathcal{H}_b(U)$ which is bounded in the natural topology is relatively compact in the compact-open topology τ_o.

QUESTION: Is this true for other topologies, such as τ_∞, τ_ω?

PROOF OF THE CARTAN-THULLEN THEOREM (part II):

c) $\Rightarrow$ b). If $CS_b(U)$ is of first category in $\mathcal{H}_b(U)$ then, since $\mathcal{H}_b(U)$ is a complete metric space in the natural topology, $S_b(U) \neq \emptyset$, and thus U is a $\mathcal{H}_b$-domain of existence.

b) $\Rightarrow$ a) is obvious.

a) $\Rightarrow$ c). Let V and W be non-empty connected open subsets of E such that $W \subset U \cap V$ and $V \not\subset U$. $\mathcal{H}_b(U,V,W)$ denotes the subalgebra of $\mathcal{H}_b(U)$ consisting of all functions $f \in \mathcal{H}_b(U)$ for which there exists a (necessarily unique) $g \in \mathcal{H}_b(V)$ such that $f = g$ in W. For each $m \in \mathbb{N}$, let $\mathcal{H}_{b,m}(U,V,W)$ be the convex subset of $\mathcal{H}_b(U,V,W)$ consisting of all $f \in \mathcal{H}_b(U,V,W)$ for which the corresponding $g \in \mathcal{H}_b(V)$ satisfies the relation $|g| \leq m$ in V.

We claim that $\mathcal{H}_{b,m}(U,V,W)$ is closed in $\mathcal{H}_b(U)$. Since $\mathcal{H}_b(U)$ with the natural topology is metrizable, it suffices to show that the limit of a convergent sequence in $\mathcal{H}_{b,m}(U,V,W)$ belongs to $\mathcal{H}_{b,m}(U,V,W)$. Let $\{f_j\}_{j\in\mathbb{N}}$ be a sequence in $\mathcal{H}_{b,m}(U,V,W)$, and suppose that $f_j \to f$ in $\mathcal{H}_b(U)$ as $j \to \infty$. For each $j \in \mathbb{N}$, let g_j be the corresponding element of $\mathcal{H}_b(V)$ such that $f_j = g_j$ in W. Since $|g_j| \leq m$ in V for every $j \in \mathbb{N}$, it follows from Montel's Theorem that $\{g_j\}_{j\in\mathbb{N}}$ has a subsequence which converges to a $g \in \mathcal{H}_b(V)$ uniformly on the compact subsets of V. In particular, $g_j \to g$ pointwise in V, and since $f_j = g_j$ in W, it follows that $f = g$ in W. Since $|g_j| \leq m$ in V for every $j \in \mathbb{N}$, $|g| \leq m$ in V. Therefore $f \in \mathcal{H}_{b,m}(U,V,W)$, and hence $\mathcal{H}_{b,m}(U,V,W)$ is closed in $\mathcal{H}_b(U)$ for every $m \in \mathbb{N}$.

We claim next that the complement $\complement\mathcal{H}_{b,m}(U,V,W)$ of $\mathcal{H}_{b,m}(U,V,W)$ in $\mathcal{H}_b(U)$ is dense in $\mathcal{H}_b(U)$, in other words, $\mathcal{H}_{b,m}(U,V,W)$ is nowhere dense in $\mathcal{H}_b(U)$. Since $\complement\mathcal{H}_b(U,V,W) \subset \complement\mathcal{H}_{b,m}(U,V,W)$, it will suffice to prove that $\complement\mathcal{H}_b(U,V,W)$ is dense in $\mathcal{H}_b(U)$. But this follows from the fact that U is a domain of $\mathcal{H}_b$-holomorphy, since then $\mathcal{H}_b(U,V,W)$ is a proper subspace of $\mathcal{H}_b(U)$. (The complement $\complement G$ in a topological vector space H of a proper subspace G is always a dense subset of H. To see this, note first that it suffices to prove that G lies in the closure of $\complement G$. If $b \in G$, let $a \in \complement G$. Then $b+\lambda a \to b$ as $\lambda \to 0$, $\lambda \in \mathbb{C}$, $\lambda \neq 0$, and $b+\lambda a \in \complement G$ for $\lambda \in \mathbb{C}$, $\lambda \neq 0$. Hence b lies in the closure of $\complement G$.)

Finally, we show that $\complement S_b(U)$ is the union of a countable family of nowhere dense sets of the form $\mathcal{H}_{b,m}(U,V,W)$.

Let M be a countable dense subset of E. If $f \in \mathcal{CS}_b(U)$ then $f \in \mathcal{H}_b(U,V,W)$ for some V, W. Let $g \in \mathcal{H}_b(V)$ be such that $f = g$ in W, and let W_o be the connected component of $U \cap V$ containing W. From the proof of Proposition 12.2, there exists $\xi \in V \cap \partial U \cap \partial W_o$. Let $r > 0$ be the distance of ξ from ∂V. Then $B_r(\xi) \subset V$ and $B_r(\xi) \not\subset U$. Now choose a point $\eta \in M \cap W_o$ sufficiently close to ξ, and a rational number s sufficiently close to r so that the ball $V' = B_s(\eta)$ is contained in V, $V' \not\subset U$ and $\sup_{V'} |g| < \infty$. Let $m \in \mathbb{N}$ be such that $\sup_{V'} |g| \leq m$, and let t be a sufficiently small positive rational number so that $B_t(\eta) \subset W_o$. Let $W' = B_t(\eta)$ and $g' = g|_{W'}$. Then $f = g'$ in W', and so $f \in \mathcal{H}_{b,m}(U,V',W')$. Since the family of sets $\mathcal{H}_{b,m}(U,V',W')$ defined in this way is countable, $\mathcal{CS}_b(U)$ is the union of a countable family of nowhere dense sets. Therefore $\mathcal{CS}_b(U)$ is of first category in $\mathcal{H}_b(U)$. Q.E.D.

REMARK 13.5: There are various open questions relating to the above. For example, what are the complex Banach spaces for which the Cartan-Thullen theorem holds? Also open is the same question with $\mathcal{H}(U)$ in place of $\mathcal{H}_b(U)$. The answer to the last question is yes when the Levi problem has solution, for example, when E is a Banach space with a Schauder basis. See Dineen [L].

PART II

THE LOCALLY CONVEX CASE

CHAPTER 14

NOTATION AND MULTILINEAR MAPPINGS

Unless stated otherwise, E and F will denote complex locally convex spaces, and U will denote a non-empty open subset of E. $\mathbb{N}$, $\mathbb{R}$ and $\mathbb{C}$ denote respectively the sets of natural numbers, of real numbers and of complex numbers. $\mathbb{N}^*$ denotes the set $\{1,2,3,\ldots\}$.

$SC(E)$ and $SC(F)$ denote respectively the sets of continuous seminorms on E and F.

DEFINITION 14.1 Let $m \in \mathbb{N}^*$. $\mathcal{L}_a(^mE;F)$ denotes the set of all m-linear mappings of $E^m = E\times E \times\ldots\times E$ (m times) into F, the operations of addition and scalar multiplication being defined pointwise. $\mathcal{L}_{as}(^mE;F)$ denotes the subspace of $\mathcal{L}_a(^mE;F)$ of all symmetric m-linear mappings. Thus $A \in \mathcal{L}_{as}(^mE;F)$ means that

$$A(x_{\sigma(1)},\ldots,x_{\sigma(m)}) = A(x_1,\ldots,x_m)$$

for every $x_1,\ldots,x_m \in E$ and every $\sigma \in S_m$, S_m being the set of all permutations of $\{1,2,\ldots,m\}$.

If $A \in \mathcal{L}_a(^mE;F)$, the symmetrization of A is the element A_s of $\mathcal{L}_{as}(^mE;F)$ defined by

$$A_s(x_1,x_2,\ldots,x_m) = \frac{1}{m!} \sum_{\sigma\in S_m} A(x_{\sigma(1)},x_{\sigma(2)},\ldots,x_{\sigma(m)}),$$

for $x_1, x_2, \ldots, x_m \in E$.

We denote by $\mathcal{L}(^mE;F)$ and $\mathcal{L}_s(^mE;F)$ respectively the vector subspaces of $\mathcal{L}_a(^mE;F)$ and $\mathcal{L}_{as}(^mE;F)$ consisting of continuous mappings.

For $m = 0$, we define $\mathcal{L}_a(^oE;F) := \mathcal{L}_{as}(^oE;F) :=$ $:= \mathcal{L}(^oE;F) := \mathcal{L}_s(^oE;F) := F$ as vector spaces, and we set $A_s = A$ for $A \in \mathcal{L}_a(^oE;F)$.

It is easy to see that the mapping $A \mapsto A_s$ is a projection of $\mathcal{L}_a(^mE;F)$ onto $\mathcal{L}_{as}(^mE;F)$ which maps $\mathcal{L}(^mE;F)$ onto $\mathcal{L}_s(^mE;F)$ for every $m \in \mathbb{N}$. In the case $F = \mathbb{C}$ we write for simplicity, $\mathcal{L}_a(^mE;\mathbb{C}) = \mathcal{L}_a(^mE)$, $\mathcal{L}_{as}(^mE;\mathbb{C}) = \mathcal{L}_{as}(^mE)$, $\mathcal{L}(^mE;\mathbb{C}) = \mathcal{L}(^mE)$ and $\mathcal{L}_s(^mE;\mathbb{C}) = \mathcal{L}_s(^mE)$. For $m = 1$, we write $\mathcal{L}_a(^1E;F) = \mathcal{L}_a(E;F)$ and $\mathcal{L}(^1E;F) = \mathcal{L}(E;F)$. If $E = \mathbb{C}$, the spaces $\mathcal{L}_a(^m\mathbb{C};F)$, $\mathcal{L}_{as}(^m\mathbb{C};F)$, $\mathcal{L}(^m\mathbb{C};F)$ and $\mathcal{L}_s(^m\mathbb{C};F)$ are all naturally isomorphic with one another. For if $\lambda_1, \ldots, \lambda_m \in \mathbb{C}$, then $A(\lambda_1, \ldots, \lambda_m) = \lambda_1 \ldots \lambda_m A(1, \ldots, 1)$, and so the mapping $A \mapsto A(1, \ldots, 1) \in F$ is an isomorphism.

DEFINITION 14.2 Let $A \in \mathcal{L}_a(^mE;F)$ and $x \in E$. Ax^m is defined as follows: if $m = 0$, $Ax^o = A \in F$. If $m \in \mathbb{N}^*$,

$$Ax^m = A(\overbrace{x, x, \ldots, x}^{m \text{ times}}).$$

More generally, let $A \in \mathcal{L}_a(^mE;F)$, $x_1, \ldots, x_k \in E$, $k \in \mathbb{N}^*$, $m, n_1, n_2, \ldots, n_k \in \mathbb{N}$, and $n = n_1 + n_2 + \ldots + n_k \leq m$. Then $Ax_1^{n_1} \ldots x_k^{n_k}$ is defined as follows: If $m = 0$, $Ax_1^{n_1} \ldots x_k^{n_k} = A$. If $m = n > 0$,

$$Ax_1^{n_1} \ldots x_k^{n_k} = A(x_1, \ldots, x_1, x_2, \ldots, x_2, \ldots, x_k, \ldots, x_k) \in F,$$

where each x_i is repeated n_i times if $n_i > 0$, and omitted if $n_i = 0$. And if $m > n$, $Ax_1^{n_1}\cdots x_k^{n_k}$ is defined by

$$(Ax_1^{n_1}\cdots x_k^{n_k})(y_1,\ldots,y_{m-n}) = A(\overbrace{x_1,\ldots,x_1}^{n_1 \text{ times}},\ldots,\overbrace{x_k,\ldots,x_k}^{n_k \text{ times}},y_1,\ldots,y_{m-n})$$

where $y_1,\ldots,y_{m-n} \in E$.

Then $Ax_1^{n_1}\cdots x_k^{n_k} \in \mathcal{L}_a({}^{m-n}E;F)$ in each case, and $Ax_1^{n_1}\cdots x_k^{n_k}$ is symmetric if A is symmetic, and continuous if A is continuous.

LEMMA 14.1 (Newton's Formula). Let $A \in \mathcal{L}_{as}({}^mE;F)$, $x_1,\ldots,x_k \in E$, $k \in \mathbb{N}^*$, $m,n \in \mathbb{N}$ and $n \le m$. Then

$$A(x_1+x_2+\cdots+x_k)^n = \Sigma \frac{n!}{n_1!\cdots n_k!} Ax_1^{n_1}\cdots x_k^{n_k},$$

the sum being taken over all $n_1,\ldots,n_k \in \mathbb{N}$ for which $n = n_1 +\cdots+ n_k$.

PROOF: The case $n = 0$ is trivial. If $m = n > 0$, then $A(x_1+\cdots+x_k)^n = A(x_1+\cdots+x_k,\ldots,x_1+\cdots+x_k)$. If $m > n > 0$, then for each $(y_1,\ldots,y_{m-n}) \in E^{m-n}$,

$$A(x_1+\cdots+x_k)^n(y_1,\ldots,y_{m-n}) =$$

$$A(\overbrace{x_1+\cdots+x_k,\ldots,x_1+\cdots+x_k}^{n \text{ times}},y_1,\ldots,y_{m-n}).$$

In each case, the given expression may be expanded, using the fact that A is multilinear and symmetric, and it is easy to see that the number of occurrences of $Ax_1^{n_1}\cdots x_k^{n_k}$ is equal to the number of permutations of $x_1,\ldots,x_k$, where x_1 is repeated n_1 times,...,x_k is repeated n_k times. Since this number is $\frac{n!}{n_1!\ldots!n_k!}$ the lemma is proved. Q.E.D.

CHAPTER 15

POLYNOMIALS

DEFINITION 15.1 A mapping $P: E \to F$ is an m-homogeneous polynomial, where $m \in \mathbb{N}$, if there exists $A \in \mathcal{L}_a({}^mE;F)$ such that

$$P(x) = Ax^m \quad \text{for every} \quad x \in E.$$

To express this relationship between P and A, we write $P = \hat{A}$. It is easy to see that $\hat{A}_s = \hat{A}$. If $m \geq 1$, and $\Delta: E \to E^m$ is the diagonal mapping, $\Delta(x) = (x,\ldots,x)$ for each $x \in E$, then $P = A \circ \Delta$. If P is an m-homogeneous polynomial, we have

$$P(\lambda x) = \lambda^m P(x) \quad \text{for} \quad x \in E, \ \lambda \in \mathbb{C}.$$

We denote by $\mathcal{P}_a({}^mE;F)$ the vector space of all m-homogeneous polynomials from E into F, the vector operations being defined pointwise. $\mathcal{P}({}^mE;F)$ denotes the subspace of $\mathcal{P}_a({}^mE;F)$ of continuous m-homogeneous polynomials. In the case $m = 0$, we take $\mathcal{P}_a({}^0E;F) = \mathcal{P}({}^0E;F)$ to be the vector space of constant mappings of E into F, which is naturally isomorphic with the vector space F. When $F = \mathbb{C}$, we write $\mathcal{P}_a({}^mE;\mathbb{C}) = \mathcal{P}_a({}^mE)$ and $\mathcal{P}({}^mE;\mathbb{C}) = \mathcal{P}({}^mE)$. When $E = \mathbb{C}$, $\mathcal{P}_a({}^m\mathbb{C};F)$ and $\mathcal{P}({}^m\mathbb{C};F)$ are both naturally isomorphic with F as vector spaces.

REMARK 15.1: If $E_1,\dots,E_m$ and F are complex locally convex spaces, $m \in \mathbb{N}^*$, then the vector spaces $\mathcal{L}_a(E_1,\dots,E_m;F)$ and $\mathcal{L}(E_1,\dots,E_m;F)$ of m-linear mappings and of continuous m-linear mappings of $E_1 \times\dots\times E_m$ into F, respectively, are subspaces of $\mathcal{P}_a(^m(E_1 \times\dots\times E_m);F)$ and $\mathcal{P}(^m(E_1 \times\dots\times E_m);F)$ respectively. To see this, let $A \in \mathcal{L}_a(^mE_1,\dots,E_m;F)$. If $X_1 = (x_1^1,\dots,x_1^m)$, $X_2 = (x_2^1,\dots,x_2^m)$ $,\dots,$ $X_m = (x_m^1,\dots,x_m^m) \in E_1 \times\dots\times E_m$, define $B\colon (E_1 \times\dots\times E_m)^m \to F$ by

$$B(X_1,\dots,X_m) = A(x_1^1,x_1^2,\dots,x_m^m).$$

It is easy to see that $B \in \mathcal{L}_a(^m(E_1 \times\dots\times E_m);F)$, and that if $X_1 = X_2 =\dots= X_m = X = (x_1,\dots,x_m)$, then

$$\hat{B}(X) = BX^m = A(x_1,x_2,\dots,x_m)$$

for every $(x_1,\dots,x_m) \in E_1 \times\dots\times E_m$. Thus statements concerning polynomials may also be applied to multilinear mappings.

LEMMA 15. 1 (The Polarization Formula). If $A \in \mathcal{L}_{as}(^mE;F)$, $m \in \mathbb{N}^*$, and $x_1,\dots,x_m \in E$, then

$$A(x_1,x_2,\dots,x_m) = \frac{1}{m!\,2^m} \Sigma\, \varepsilon_1\varepsilon_2\dots\varepsilon_m\, \hat{A}(\varepsilon_1x_1+\varepsilon_2x_2+\dots+\varepsilon_mx_m),$$

the summation extending over all possible values of $\varepsilon_1 = \pm 1,\dots,\varepsilon_m = \pm 1$.

PROPOSITION 15.1 The mapping

$$A \in \mathcal{L}_a(^mE;F) \longmapsto \hat{A} \in \mathcal{P}_a(^mE;F)$$

is linear, and induces an isomorphism between the vector spaces $\mathcal{L}_{as}(^mE;F)$ and $\mathcal{P}_a(^mE;F)$, and between the vector spaces $\mathcal{L}_s(^mE;F)$ and $\mathcal{P}_a(^mE;F)$, for every $m \in \mathbb{N}$.

PROOF: The case $m = 0$ is trivial. Let $m \geq 1$. The mapping $A \in \mathcal{L}_a(^mE;F) \longmapsto \hat{A} \in \mathcal{L}_a(^mE;F)$ is easily seen to be linear and surjective, from the definition of $\mathcal{P}_a(^mE;F)$. Since $\hat{A}_s = \hat{A}$, the mapping induces a surjective mapping of $\mathcal{L}_{as}(^mE;F)$ onto $\mathcal{P}_a(^mE;F)$. The polarization formula shows that this mapping is injective and maps $\mathcal{L}_s(^mE;F)$ onto $\mathcal{P}(^mE;F)$. Q.E.D.

DEFINITION 15.2 Let $\mathcal{F}_a(E;F)$ and $\mathcal{F}(E;F)$ denote respectively the vector space of all mappings of E into F and the vector space of all continuous mappings of E into F. An element of the algebraic sum of the subspaces $\mathcal{P}_a(^mE;F)$ of $\mathcal{F}_a(E;F)$, $m \in \mathbb{N}$, is called a polynomial from E into F. Thus, a mapping $P: E \to F$ is a polynomial if there exist $m \in \mathbb{N}$ and $P_k \in \mathcal{P}_a(^kE;F)$, $k = 0,1,\ldots,m$, such that

$$P = P_o + P_1 + \ldots + P_m .$$

$\mathcal{P}_a(E;F)$ denotes the vector space of polynomials from E into F, and $\mathcal{P}(E;F)$ denotes the subspace of continuous polynomials. When $F = \mathbb{C}$, we write $\mathcal{P}_a(E;C) = \mathcal{P}_a(E)$ and $\mathcal{P}(E;\mathbb{C}) = \mathcal{P}(E)$.

PROPOSITION 15.2 $\mathcal{P}_a(E;F)$ and $\mathcal{P}(E;F)$ are the direct sums of the families $\{\mathcal{P}_a(^mE;F)\}_{m\in\mathbb{N}}$ and $\{\mathcal{P}(^mE;F)\}_{m\in\mathbb{N}}$ respectively.

PROOF: We show first that the family of subspaces $\mathcal{P}_a(^m E;F)$ of $\mathcal{F}_a(E;F)$, $m \in \mathbb{N}$, is linearly independent. Thus if $P_k \in \mathcal{P}_a(^k E;F)$, $k = 0,1,\ldots,m$, $m \in \mathbb{N}$, and

(1) $$P_o + P_1 + \ldots + P_m = 0,$$

we must show that $P_o = P_1 = \ldots = P_m = 0$. We prove this by induction. The assertion is trivial if $m = 0$; we assume its truth for $m-1$, $m \geq 1$. Now condition (1) implies that

(2) $$\lambda_m \sum_{k=0}^{m} P_k(x) = 0 \qquad \text{and}$$

(3) $$\sum_{k=0}^{m} \lambda^k P_k(x) = 0$$

for every $x \in E$, $\lambda \in \mathbb{C}$. Subtracting (3) from (2), we have

(4) $$(\lambda^m - 1)P_o(x) + \ldots + (\lambda^m - \lambda^{m-1})P_{m-1}(x) = 0$$

for every $x \in E$, $\lambda \in \mathbb{C}$. If we choose $\lambda \in \mathbb{C}$ so that $\lambda^m \neq \lambda^k$ for $k = 0,1,\ldots,m-1$, then our induction hypothesis applied to (4) yields

$$P_o = P_1 = \ldots = P_{m-1} = 0.$$

And from (1) we have $P_m = 0$.

To show that $\mathcal{P}(E;F)$ is the direct sum of the family $\mathcal{P}(^m E;F)$, $m \in N$, let $P \in \mathcal{P}(E;F)$. Then there exist $P_k \in \mathcal{P}_a(^k E;F)$, $k = 0,1,\ldots,m$, $m \in N$ such that

(a) $P = P_o + P_1 + \ldots + P_m$.

We must show that $P_k \in \mathcal{P}(^k E;F)$, $k = 0,1,\ldots,m$. This is trivial for $m = 0$; we assume the truth of this assertion for $m-1$. From (a) we obtain

$$\text{(b)} \quad \lambda^m P(x) - P(\lambda x) = (\lambda^m - 1)P_o(x) + \dots + (\lambda^m - \lambda^{m-1})P_{m-1}(x)$$

for every $x \in E$, $\lambda \in \mathbb{C}$. Fix $\lambda \in \mathbb{C}$ such that $\lambda^m \neq \lambda^k$ for $k = 0,1,\dots,m-1$. Since the left hand side of (b) is a continuous function of x, it follows by the induction hypothesis that $P_o, P_1, \dots, P_{m-1}$ are continuous. Hence, from (a), P_m is also continuous. Q.E.D.

REMARK 15.2: The preceeding proof shows that the following is true: if $m \in \mathbb{N}$, $P = P_o + \dots + P_m$, with $P_k \in \mathcal{P}_a({}^kE;F)$ for $k = 0,1,\dots,m$, then $P \in \mathcal{P}(E;F)$ if and only if $P_k \in \mathcal{P}({}^kE;F)$ for every k, $k = 0,1,\dots,m$.

DEFINITION 15.3 Let $P \in \mathcal{P}_a(E;F)$, $P \neq 0$. Then there is a unique $m \in \mathbb{N}$ and unique polynomials $P_o, P_1, \dots, P_m$, with $P_k \in \mathcal{P}_a({}^kE;F)$, $k = 0,1,\dots,m$, such that $P = P_o + \dots + P_m$ and $P_m \neq 0$. The non-negative integer m is called the degree of P. The polynomial $P = 0$ is conventionally assigned a degree of either -1 or $-\infty$, depending on the context.

REMARK 15.3: Let E, F and G be complex vector spaces, $P \in \mathcal{P}_a(E;F)$ and $Q \in \mathcal{P}_a(E;G)$. Then $Q \circ P \in \mathcal{P}_a(E;G)$. If $P \neq 0$ and $Q \neq 0$, the degree of $Q \circ P$ is less than or equal to the product of the degrees of Q and P.

DEFINITION 15.4 A mapping $f: U \to F$ is said to be amply bounded if for every $\xi \in U$ and every $\beta \in CS(F)$ there is a neighbourhood V of ξ contained in U such that $\sup_{x \in V} \beta\{f(x)\} < \infty$.

PROPOSITION 15.3 For $P \in \mathcal{P}_a(E;F)$, the following are equivalent:

(1) P is continuous.

(2) P is amply bounded.

(3) P is continuous at one point.

(4) P is amply bounded at one point.

PROOF: The implications (1) $\Rightarrow$ (2) $\Rightarrow$ (4) and (1) $\Rightarrow$ (3) $\Rightarrow$ (4) are clear. Thus it suffices to prove (4) $\Rightarrow$ (1). Let $\beta \in CS(F)$ and $\xi \in E$. By hypothesis, there is a non-empty open subset U of E and a number $M \geq 0$ such that

(a) $\beta\{P(x)\} \leq M$ for every $x \in U$.

Let $u_o \in U$, and consider the translation $t: E \to E$, $t(x) = x-u_o$. Then (a) is equivalent to

(b) $\beta\{Q(y)\} \leq M$ for every $y \in V = t(U)$,

where V is a neighbourhood of zero and $Q = P \circ t^{-1}$ is a polynomial. Since $P = Q \circ t$, continuity of P is equivalent to continuity of Q. We claim that

(c) If $Q = \sum_{j=0}^{m} Q_j$, where $Q_j \in \mathcal{P}_a({}^jE;F)$, $j = 0,1,\ldots,m$, $Q_m \neq 0$, then $\beta \circ Q$ is bounded on V if and only if $\beta \circ Q_j$ is bounded on V for $j = 0,1,\ldots,m$.

We prove this by induction. The case $m = 0$ is trivial. Assuming the truth of (c) for $m-1$, with $m \geq 1$, let $Q = \sum_{j=0}^{m} Q_j$, $Q_j \in \mathcal{P}_a({}^jE;F)$, $Q_m \neq 0$, and suppose Q is bounded on V. We have

(d) $\lambda^m Q(x) - Q(\lambda x) = \sum_{j=0}^{m-1} (\lambda^m - \lambda^j) Q_j(x)$

for every $x \in E$ and $\lambda \in \mathbb{C}$. Fixing $\lambda \in \mathbb{C}$ such that $\lambda^m - \lambda^j \neq 0$ for $j = 0,\ldots,m-1$, the polynomial $\sum_{j=0}^{m-1} (\lambda^m - \lambda^j)Q_j$, which has degree at most $m-1$, is such that $\beta\{\sum_{j=0}^{m-1} (\lambda^m - \lambda^j)Q_j\}$ is bounded on V, since $\beta\{\lambda^m Q(x) - Q(\lambda x)\}$ is bounded on V. Hence, by the induction hypothesis, $\beta \circ Q_o, \ldots, \beta \circ Q_{m-1}$ are bounded on V. Therefore $\beta \circ Q_m$ is also bounded on V, since

$$\beta \circ Q_m(x) \leq \beta \circ Q(x) + \beta\{\sum_{j=0}^{m-1} Q_j(x)\}.$$

This completes the proof of (c).

Now let $A_j \in \mathcal{L}_{as}({}^jE;F)$ be such that $Q_j = \hat{A}_j$, $j = 0,1,\ldots,m$. Then

$$\text{(c)} \qquad Q(x) = \sum_{j=0}^{m} A_j x^j.$$

For $j = 1,2,\ldots,m$, let W_j be a balanced neighbourhood of the origin in E such that

$$\overbrace{W_j + \ldots + W_j}^{j \text{ times}} \subset V.$$

Then for $(x_1,\ldots,x_j) \in (W_j)^j$, we have $\sum_{i=1}^{j} \epsilon_i x_i \in V$ if $\epsilon_i = \pm 1$ for $i = 1,\ldots,j$. It follows from the polarization formula that each $\beta \circ A_j$ is bounded on the neighbourhood $(W_j)^j$ of the origin in E^j, for $j = 1,2,\ldots,m$. Hence each A_j is continuous on E^j. But A_o is continuous on E and so by (c), Q is continuous on E. Q.E.D.

CHAPTER 16

TOPOLOGIES ON SPACES OF MULTILINEAR MAPPINGS AND HOMOGENEOUS POLYNOMIALS

DEFINITION 16.1 Let E and F be complex seminormed spaces, and $m \in \mathbb{N}^*$. Then a seminorm is defined in a natural way on $\mathcal{L}_s({}^mE;F)$ by defining

$$\|A\| = \sup \frac{\|A(x_1,\ldots,x_m)\|}{\|x_1\|\ldots\|x_m\|}$$

for $A \in \mathcal{L}_s({}^mE;F)$, where the supremum is taken over all $x_1,\ldots,x_m \in E$ such that $\|x_1\| \neq 0,\ldots,\|x_m\| \neq 0$; if the seminorm of E is identically zero, we define $\|A\| = 0$. We then have

$$\|A\| = \sup\{\|A(x_1,\ldots,x_m)\| : \|x_1\| \leq 1,\ldots,\|x_m\| \leq 1\}$$

and

$$\|A(x_1,\ldots,x_m)\| \leq \|A\| \ \|x_1\|\ldots\|x_m\| \quad \text{for all} \quad x_1,\ldots,x_m \in E.$$

For $m = 0$, $\mathcal{L}_s({}^oE;F)$ is seminormed in a natural way by the seminorm of F.

The natural seminorm of $\mathcal{P}({}^mE;F)$, $m \in \mathbb{N}^*$, is defined in a similar way; for $P \in \mathcal{P}({}^mE;F)$, we define

$$\|P\| = \sup\{\frac{\|P(x)\|}{\|x\|^m} : x \in E, \ \|x\| \neq 0\}.$$

If the seminorm of E is identically zero, we define $\|P\| = 0$. For $m = 0$, $\mathcal{P}({}^oE;F)$ is seminormed in a natural way by the seminorm of F. If F is a normed space or a Banach space, then $\mathcal{L}_s(^mE;F)$ and $\mathcal{P}(^mE;F)$ are normed spaces or Banach spaces respectively, for every $m \in \mathbb{N}$.

We point out that the same symbol, $\| \ \|$, is used to denote the seminorms of E, F, $\mathcal{L}_s(^mE;F)$ and $\mathcal{P}(^mE;F)$.

PROPOSITION 16.1 Let E and F be complex seminormed spaces, and $m \in \mathbb{N}$. The mapping $A \in \mathcal{L}_s(^mE;F) \mapsto \hat{A} \in \mathcal{P}(^mE;F)$ is an isomorphism of vector spaces, and a homeomorphism. Furthermore,

$$(*) \qquad \|\hat{A}\| \le \|A\| \le \frac{m^m}{m!} \|\hat{A}\|$$

for every $A \in \mathcal{L}_s(^mE;F)$.

PROOF: The case $m = 0$ is trivial, m^m being defined to be 1 for convenience. Let $m > 0$. We know, by Proposition 15.1 that the mapping $A \mapsto \hat{A}$ is an isomorphism of vector spaces. If we prove the inequality (*), it will follow that this mapping is a homeomorphism. It is easy to see that $\|\hat{A}\| \le \|A\|$. If the seminorm of E is identically zero, the proposition is trivial. Suppose then that the seminorm of E is not identically zero. For all $x_1,\ldots,x_m \in E$ such that $\|x_1\| = \ldots = \|x_m\| = 1$, we have

$$\|\hat{A}(\varepsilon_1 x_1 + \ldots + \varepsilon_m x_m)\| \le \|\hat{A}\| m^m \quad \text{if} \quad \varepsilon_1 = \pm 1, \ldots, \varepsilon_m = \pm 1.$$

Using the polarization formula (Proposition 2.1), we have

$$\|A(x_1,\ldots,x_m)\| \le \frac{1}{m!\,2^m} 2^m \|\hat{A}\| m^m = \frac{m^m}{m!} \|\hat{A}\|.$$

Therefore $\|A\| \le \frac{m^m}{m!} \|\hat{A}\|$. Q.E.D.

REMARK 16.1: In general, the mapping $A \in \mathcal{L}_s(^mE;F) \mapsto \hat{A} \in \mathcal{P}(^mE;F)$ does not preserve the seminorms. In fact, for each $m \in \mathbb{N}$, the smallest constant $C_m \geq 0$ such that $\|A\| \leq C_m\|\hat{A}\|$ for $A \in \mathcal{L}_s(^mE;F)$, independently of E and F, is $m^m/m!$ This is shown by the following example.

EXAMPLE 16.1 Let e be ℓ^1, the Banach space of all sequences $x = (x_1,x_2,\ldots)$ of complex numbers such that $\sum_{n=1}^{\infty} |x_n| < \infty$, with the norm $\|x\| = \sum_{n=1}^{\infty} |x_n|$. If $A_m \in \mathcal{L}_s(^mE)$ is defined by

$$A_m(x^1,x^2,\ldots,x^m) = \frac{1}{m!} \sum_{\sigma \in S_m} x_1^{\sigma(1)}.x_2^{\sigma(2)}\ldots x_m^{\sigma(m)},$$

then $\|A_m\| = 1/m!$ and $\|\hat{A}_m\| = 1/m^m$. Thus $\|A_m\| = \frac{m^m}{m!}\|\hat{A}_m\|$. We refer the reader to Part I, Example 1.2 for details.

REMARK 16.2: For $\alpha \in CS(E)$, E_α denotes the space E seminormed by α. If $\alpha \in CS(E)$ and $\beta \in CS(F)$, $\mathcal{L}_s(^mE_\alpha;F_\beta)$ and $\mathcal{P}(^mE_\alpha;F_\beta)$ denote respectively the vector spaces of continuous symmetric m-linear mappings of E_α^m into F_β, and of continuous m-homogeneous polynomials from E_α into F_β. For $A \in \mathcal{L}_s(^mE_\alpha;F_\beta)$, $P \in \mathcal{P}(^mE_\alpha;F_\beta)$, $\|A\|_{\alpha,\beta}$ and $\|P\|_{\alpha,\beta}$ respectively denote the natural seminorms of A and P.

REMARK 16.3: Let E and F be complex seminormed vector spaces, $m \in \mathbb{N}$, and $A \in \mathcal{L}(^mE;F)$. If $k \in \mathbb{N}^*$, $n_1,\ldots,n_k \in \mathbb{N}$ and $n_1 +\ldots+ n_k \leq m$, we define

$$\|A^{(n_1,\ldots,n_k)}\| = \sup \frac{\|Ax_1^{n_1}\ldots x_k^{n_k}\|}{\|x_1\|^{n_1}\ldots\|x_k\|^{n_k}}$$

where the supremum is taken over all $x_1,\ldots,x_k \in E$ such that

$\|x_1\| \neq 0, \ldots, \|x_k\| \neq 0$. If the seminorm of E is identically zero, we set $\|A^{(n_1,\ldots,n_k)}\| = 0$.

We then have

$$\|Ax_1^{n_1}\ldots x_k^{n_k}\| \leq \|A^{(n_1,\ldots,n_k)}\| \cdot \|x_1\|^{n_1}\ldots\|x_k\|^{n_k}.$$

In particular, we have

$$\|A^{(1,\ldots,1)}\| = \|A\| \quad \text{and} \quad \|A^{(m)}\| = \|\hat{A}\|,$$

where $m > 0$, and the number 1 is repeated m times in the first equation.

DEFINITION 16.2 Let E and F be complex locally convex spaces and $m \in \mathbb{N}$. For $\alpha \in CS(E)$ and $\beta \in CS(F)$, the seminormed spaces $\mathcal{L}_s({}^mE_\alpha;F_\beta)$ and $\mathcal{P}({}^mE_\alpha;F_\beta)$ are homeomorphic by the natural mapping $A \mapsto \hat{A}$. Therefore the locally convex spaces

$$\mathcal{L}_s({}^mE;F_\beta) = \bigcup_{\alpha\in CS(E)} \mathcal{L}_s({}^mE_\alpha;F_\beta)$$

and

$$\mathcal{P}({}^mE;F_\beta) = \bigcup_{\alpha\in CS(E)} \mathcal{P}({}^mE_\alpha;F_\beta),$$

with their corresponding locally convex inductive topologies, are homeomorphic by the natural mapping $A \mapsto \hat{A}$.

Finally, consider the locally convex spaces

$$\mathcal{L}_s({}^mE;F) = \bigcap_{\beta\in CS(F)} \mathcal{L}_s({}^mE;F_\beta)$$

and

$$\mathcal{P}({}^mE;F) = \bigcap_{\beta\in CS(F)} \mathcal{P}({}^mE;F_\beta),$$

with the corresponding projective topologies; the natural

isomorphism $A \mapsto \hat{A}$ establishes a homeomorphism between these spaces. The locally convex topologies obtained in this way are known as the limit topologies on $\mathcal{L}_s(^mE;F)$ and $\mathcal{P}(^mE;F)$.

DEFINITION 16.3 The bounded topologies on $\mathcal{L}_s(^mE;F)$ and $\mathcal{P}(^mE;F)$ are defined respectively by the seminorms

$$A \in \mathcal{L}_s(^mE;F) \mapsto \sup\{\beta[A(x_1,\dots,x_m)] : x_1 \in X_1,\dots,x_m \in X_m\}$$

$$P \in \mathcal{P}(^mE;F) \mapsto \sup\{\beta[P(x)] : x \in X\},$$

where $\beta \in CS(F)$ and $X_1,\dots,X_m$ and X are bounded subsets of E. In the case $m = 0$, these seminorms are defined to be $\beta(A)$ and $\beta(P)$ respectively.

The compact topologies on $\mathcal{L}_s(^mE;F)$ and $\mathcal{P}(^mE;F)$ are defined respectively by the seminorms

$$A \in \mathcal{L}_s(^mE;F) \mapsto \sup\{\beta[A(x_1,\dots,x_m)] : x_1 \in X_1,\dots,x_m \in X_m\}$$

$$P \in \mathcal{P}(^mE;F) \mapsto \sup\{\beta[P(x) : x \in X\}$$

where $\beta \in CS(F)$, and $X_1,\dots,X_m$ and X are compact subsets of E. Again, for $m = 0$, these seminorms are $\beta(A)$ and $\beta(P)$ respectively.

The finite topologies on $\mathcal{L}_s(^mE;F)$ and $\mathcal{P}(^mE;F)$ are defined by families of seminorms defined in the same manner as above with $X_1,\dots,X_m$ and X ranging over the finite subsets of E.

It is easy to see that in the case of each of these four topologies, the locally convex spaces $\mathcal{L}_s(^mE;F)$ and $\mathcal{P}(^mE;F)$ are homeomorphic by the natural isomorphism $A \mapsto \hat{A}$.

REMARK 16.3: The four topologies which we have defined on $\mathcal{L}_s(^mE;F)$ and $\mathcal{P}(^mE;F)$ are related in the following way: limit topology $\geq$ bounded topology $\geq$ compact topology $\geq$ $\geq$ finite topology.

REMARK 16.4: The limit, bounded, compact and finite topologies can be defined in a similar way on the space $\mathcal{L}(^mE;F)$, $m \in \mathbb{N}$. Each one of these topologies induces the corresponding topology on $\mathcal{L}_s(^mE;F)$, and the natural mappings

$$\mathcal{L}(^mE;F) \to \mathcal{L}_s(^mE;F) \quad \text{and} \quad \mathcal{L}(^mE;F) \to \mathcal{P}(^mE;F)$$

are continuous.

REMARK 16.5: If $E_1,\ldots,E_m$ are locally convex spaces, $m \in \mathbb{N}^*$, we can define on the space $\mathcal{L}(E_1,\ldots,E_m;F)$ of continuous m-linear mappings of $E_1 \times\ldots\times E_m$ into F, the limit, bounded, compact and finite topologies. Each of these topologies coincides with the topology induced on $\mathcal{L}(^mE_1,\ldots,E_m;F)$ by the corresponding topology on $\mathcal{P}(^m(E_1\times\ldots\times E_m);F)$, and in each case, the natural projection of $\mathcal{P}(^m(E_1\times\ldots\times E_m);F)$ onto $\mathcal{L}(E_1,\ldots,E_m;F)$ is continuous.

CHAPTER 17

FORMAL POWER SERIES

DEFINITION 17.1 A formal power series, or more simply, a power series, from E into F about a point ξ of E is a series in the variable $x \in E$ of the form

$$\sum_{m=0}^{\infty} A_m(x-\xi)^m,$$

where $A_m \in \mathcal{L}_{as}(^mE;F)$, $m \in \mathbb{N}$. Equivalently, this series can be written in the form

$$\sum_{m=0}^{\infty} P_m(x-\xi),$$

where $P_m = \hat{A}_m$. A_m and P_m are both referred to as the coefficient of order m of the series, and ξ as the origin of the series. The space $F_a[[E]]$ of formal power series from E into F about ξ is a vector space which is canonically isomorphic to

$$\prod_{m\in\mathbb{N}} \mathcal{L}_{as}(^mE;F), \quad \text{and to} \quad \prod_{m\in\mathbb{N}} \mathcal{P}_a(^mE;F).$$

We denote by $F[[E]]$ the subspace of $F_a[[E]]$ consisting of all formal power series from E into F about ξ which are continuous on E, by which we mean that each coefficient is a continuous mapping. Then $F[[E]]$ is canonically isomorphic to

$$\prod_{m\in\mathbb{N}} \mathcal{L}_s({}^m E;F) \quad \text{and to} \quad \prod_{m\in\mathbb{N}} \mathcal{P}({}^m E;F).$$

LEMMA 17.1 Let $\xi \in E$, $P_m \in \mathcal{P}_a({}^m E;F)$ $(m \in \mathbb{N})$ and $\beta \in cs(F)$. Then

$$\lim_{m\to\infty} \beta[\sum_{k=0}^{m} P_k(x-\xi)] = 0$$

for every x in some neighbourhood of ξ if and only if $\beta \circ P_m = 0$ for every $m \in \mathbb{N}$.

PROOF: We shall prove that if $\lim_{m\to\infty} \beta[\sum_{k=0}^{m} P_k(x-\xi)] = 0$, for every x in some neighbourhood of ξ, then $\beta \circ P_m = 0$ for every m; the converse is clear.

We claim that if $u_m \in F$ $(m \in \mathbb{N})$ and $\delta > 0$, and if

a) $\lim_{m\to\infty} \beta[\sum_{k=0}^{m} \lambda^k u_k] = 0$ for $\lambda \in \mathbb{C}$, $|\lambda| \leq \delta$, then $\beta(u_m) = 0$ for every $m \in \mathbb{N}$. We prove this by induction. Taking $\lambda = 0$, we obtain $\beta(u_o) = 0$. Suppose that $\beta(u_o) = \ldots = \beta(u_{n-1}) = 0$, $n \geq 1$. It follows from a), with $\lambda = \delta$, that there exists a $c \geq 0$ such that $\beta(u_m) \leq c/\delta^m$ for every $m \in \mathbb{N}$. Applying the induction hypothesis to a),

$$\lim_{m\to\infty} \{\beta(\lambda^n u_n) - \beta(\sum_{k=n+1}^{m} \lambda^k u_k)\} = 0,$$

and hence

$$\beta(u_n) = \lim_{m\to\infty} \beta(\sum_{k=n+1}^{m} \lambda^{k-n} u_k)$$

for $\lambda \in \mathbb{C}$, $0 < |\lambda| < \delta$. This implies that

$$\beta(u_n) \leq \lim_{m\to\infty} \sum_{k=n+1}^{m} |\lambda|^{k-n} \frac{c}{\delta^k} = \frac{c|\lambda|}{\delta^n(\delta - |\lambda|)} .$$

Letting $\lambda \to 0$, we obtain $\beta(u_n) = 0$. This proves the claim.

Now let $x = \xi + \lambda t$, where $\lambda \in \mathbb{C}$ and $t \in E$. By hypothesis,

$$\lim_{m\to\infty} \beta\left(\sum_{k=0}^{m} P_k(\lambda t)\right) = \lim_{m\to\infty} \left(\sum_{k=0}^{m} \lambda^p P_k(t)\right) = 0$$

for $|\lambda| \leq \delta$, where δ is some positive real number. It follows that $\beta(P_m(t)) = 0$ for every $t \in E$, $m \in \mathbb{N}$.

Q.E.D.

CHAPTER 18

HOLOMORPHIC MAPPINGS

We adopt the Weierstrass point of view, defining holomorphy in terms of power series; one could equally well employ the Cauchy-Riemann definition in terms of complex differentiability.

DEFINITION 18.1 A mapping $f: U \to F$ is holomorphic in U if for every $\xi \in U$ there exists a sequence $A_m \in \mathcal{L}_s(^mE;F)$, $m \in \mathbb{N}$ with the following property: for every $\beta \in CS(F)$ there exists a neighbourhood V of ξ in U such that

$$\lim_{m\to\infty} \beta[f(x) - \sum_{k=0}^{m} A_k(x-\xi)^k] = 0$$

uniformly in $x \in V$.

We denote by $\mathcal{H}(U;F)$ the vector space of all holomorphic mappings of U into F, the vector operations being defined pointwise. In the case $F = \mathbb{C}$ we write $\mathcal{H}(U)$ for $\mathcal{H}(U;\mathbb{C})$.

A_m, or the polynomial $P_m = \hat{A}_m$, is called the Taylor coefficient of order m of f at ξ. If we assume that F is separated, then $A_o = f(\xi)$, and it follows from Lemma 17.1 that the sequences $\{A_m\}_{m\in\mathbb{N}}$ and $\{P_m\}_{m\in\mathbb{N}}$ are unique. We write

$$d^m f(\xi) = m!A_m, \quad \hat{d}^m f(\xi) = m!P_m = m!\hat{A}_m.$$

Either of these mappings is known as the differential of order m of f at ξ. We can then define the mappings:

$$d^m f\colon x \in U \mapsto d^m f(x) \in \mathcal{L}_s({}^m E;F)$$

$$\hat{d}^m f\colon x \in U \mapsto \hat{d}^m f(x) \in \mathcal{P}({}^m E;F), \quad m \in \mathbb{N}.$$

The series

$$\sum_{m=0}^{\infty} \frac{1}{m!} d^m f(\xi)(x-\xi)^m = \sum_{m=0}^{\infty} \frac{1}{m!} \hat{d}^m f(\xi)(x-\xi)$$

is called the Taylor series of f at ξ. For $m \in \mathbb{N}$, we write

$$\tau_{m,f,\xi}(x) = \sum_{k=0}^{m} \frac{1}{k!} d^k f(\xi)(x-\xi)^k = \sum_{k=0}^{m} \frac{1}{k!} \hat{d}^k f(\xi)(x-\xi);$$

$\tau_{m,f,\xi}$ is called the Taylor polynomial of order m of f at ξ. The difference $(f - \tau_{m,f,\xi})|_U$ is called the Taylor remainder or order m of f at ξ.

In the case $E = \mathbb{C}$, $\mathcal{L}_s({}^m E;F)$ and $\mathcal{P}({}^m E;F)$ can be identified naturally with F for each $m \in \mathbb{N}$; for $A \in \mathcal{L}_s({}^m \mathbb{C};F)$, we identify A with $A(1,1,\ldots,1) \in F$. Thus $d^m f(\xi)$ and $\hat{d}^m f(\xi)$ are identified with an element of F which we denote by $f^{(m)}(\xi)$, and refer to as the derivative of order m of f at ξ. The mapping

$$f^{(m)}\colon x \in U \mapsto f^{(m)}(x) \in F$$

is called the derivative of order m of f in U.

The mappings

$$f \in \mathcal{H}(U;F) \longmapsto d^m f \in \mathcal{F}_a(U;\mathcal{L}_s(^m E;F)) \quad \text{and}$$

$$f \in \mathcal{H}(U;F) \longmapsto \hat{d}^m f \in \mathcal{F}_a(U;\mathcal{P}(^m E;F))$$

are linear for every $m \in \mathbb{N}$.

REMARK 18.1: Suppose that the mapping $f: U \to F$ has the property that for every $\xi \in U$ there exists a sequence $A_m \in \mathcal{L}_s(^m E;F)$, $m \in \mathbb{N}$, such that

$$f(x) = \sum_{m=0}^{\infty} A_m(x-\xi)^m$$

uniformly in a neighbourhood V of ξ in U. This means that

$$\lim_{m\to\infty} \beta[f(x) - \sum_{k=0}^{m} \frac{1}{k!} d^k f(\xi)(x-\xi)^k] = 0$$

uniformly in V for every $\beta \in CS(F)$. Therefore, assuming that F is separated, the Taylor series of f at ξ represents the function f in a neighbourhood of ξ. This holds for every holomorphic mapping if E and F are, for example, normed spaces. However, it can happen that a mapping $f \in \mathcal{H}(U;F)$ cannot be uniformly represented in any neighbourhood of a point $\xi \in U$ by its Taylor series at ξ. We shall see an example of this phenomenon later (see pages 221-228).

REMARK 18.2: The concepts of holomorphicity and of the differential mapping are local in nature, in the following sense:

a) If $f \in \mathcal{H}(U;F)$ and V is non-empty open subset of U, then $f \in \mathcal{H}(V;F)$, and $d^m(f|_V) = (d^m f)|_V$, $\hat{d}^m(f|_V) = (\hat{d}^m f)|_V$ for every $m \in \mathbb{N}$.

b) If U is the union of a family $\{V_\lambda\}_{\lambda\in\Lambda}$ of non-empty open subsets, f is a mapping of U into F, and if $f \in \mathcal{H}(V_\lambda;F)$ for every $\lambda \in \Lambda$, then $f \in \mathcal{H}(U;F)$.

REMARK 18.3: Let F be separated, $f \in \mathcal{H}(U;F)$ and $\xi \in U$. If f vanishes in a neighbourhood of ξ, then $d^m(\xi) = 0$ for every $m \in \mathbb{N}$ by the uniqueness of the Taylor coefficients. Conversely, if $d^m f(\xi) = 0$ for every $m \in \mathbb{N}$, then, as we shall see later (see page 222) f vanishes in some neighbourhood of ξ. Thus every $f \in \mathcal{H}(U;F)$ is, locally, uniquely determined by its Taylor series.

REMARK 18.4: We point out that the uniqueness of the sequence $\{A_m\}_{m\in\mathbb{N}}$ of differentials in Definition 5.1 (assuming that F is separated) does not depend on the continuity of A_m, $m \in \mathbb{N}$, but only on the existence of the limits as stated in the definition.

REMARK 18.5: The definition of a holomorphic mapping between complex topological vector spaces which are not locally convex must be worded very carefully. This definition is as follows: $f\colon U \to F$ is holomorphic if for every $\xi \in U$ there exists a sequence $A_m \in \mathcal{L}_s(^mE;F)$, $m \in \mathbb{N}$, with the following property: for every neighbourhood W of zero in F there exists a neighbourhood V of ξ in U such that for every $\varepsilon > 0$ there is a corresponding $M \in \mathbb{N}$ for which

$$f(x) - \sum_{k=0}^{m} A_k(x-\xi)^k \in \varepsilon W \quad \text{for} \quad m \geq M, \quad \text{and} \quad x \in V.$$

When F is locally convex, this is clearly equivalent to Definition 5.1. However, in general, we may not omit the phrase "for every $\varepsilon > 0$", and replace εW by W, as might be expected. For, if this were done, every continuous mapping $f: U \to F$ would satisfy the definition, taking $A_o = f(\xi)$, and $A_m = 0$ for $m > 0$. It would not be desirable to have a definition which allowed every continuous mapping to be holomorphic.

CHAPTER 19

SEPARATION AND PASSAGE TO THE QUOTIENT

REMARK 19.1: If, in Definition 5.1, the locally convex space F is not assumed to be separated the sequence $\{A_m\}_{m\in\mathbb{N}}$ corresponding to $f \in \mathcal{H}(U;F)$ and $\xi \in U$ is not necessarily unique. Let $K\mathcal{L}_s({}^mE;F)$ denote the subspace of $\mathcal{L}_s({}^mE;F)$ consisting of the mappings whose image lies in the closure of $\{0\}$ in F, that is, in the set $\{y \in F : \beta(y) = 0 \text{ for every } \beta \in CS(F)\}$. Then if $D_m \in K\mathcal{L}_s({}^mE;F)$ and $\{A_m\}_{m\in\mathbb{N}}$ is a sequence of Taylor coefficients of f at ξ, the sequence $\{A_m+D_m\}_{m\in\mathbb{N}}$ is also a sequence of Taylor coefficients of f at ξ. In order that the differentials be uniquely defined, we are forced to consider $d^mf(\xi)$ as an element of the quotient space $\mathcal{L}_s({}^mE;F)/K\mathcal{L}_s({}^mE;F)$, that is, as an equivalence class modulo $K\mathcal{L}_s({}^mE;F)$. Similarly, $\hat{d}^mf(\xi)$ is an element of $\mathcal{P}({}^mE;F)/K\mathcal{P}({}^mE;F)$. This procedure applies to all cases, whether or not F is separated. We remark that $K\mathcal{L}_s({}^mE;F)$ is the closure of the origin in $\mathcal{L}_s({}^mE;F)$ for any one of the four topologies which we have defined on this space. The same is true of $K\mathcal{P}({}^mE;F)$ in $\mathcal{P}({}^mE;F)$.

REMARK 19.2: The general case can also be reduced to the case in which F is separated in the following way. Let F_s be the separated space associated to F. The canonical linear mapping $\pi: F \to F_s$ is continuous and open. It is easy to prove that $f: U \to F$ is holomorphic in U if and only if $f_s = \pi \circ f: U \to F_s$ is holomorphic. Furthermore, if $\xi \in U$ then $d^m f(\xi)$ and $d^m f_s(\xi)$ correspond under the canonical isomorphism between $\mathcal{L}_s(^m E;F)/K\mathcal{L}_s(^m E;F)$ and $\mathcal{L}_s(^m E;F_s)$. Similarly, $\hat{d}^m f(\xi)$ and $\hat{d}^m f_s(\xi)$ correspond under the canonical isomorphism between $\mathcal{P}(^m E;F)/K\mathcal{P}(^m E;F)$ and $\mathcal{P}(^m E;F_s)$.

REMARK 19.3: This reduction to separated spaces can be carried one step farther. Let $\pi_E: E \to E_s$ and $\pi_F: F \to F_s$ be the canonical linear mappings of E and F onto their respective associated separated spaces. Then $U_s = \pi_E(U)$ is a non-empty open subset of E_s, and $f: U \to F$ is holomorphic if and only if there is a holomorphic mapping $f_s: U_s \to F_s$ such that the diagram

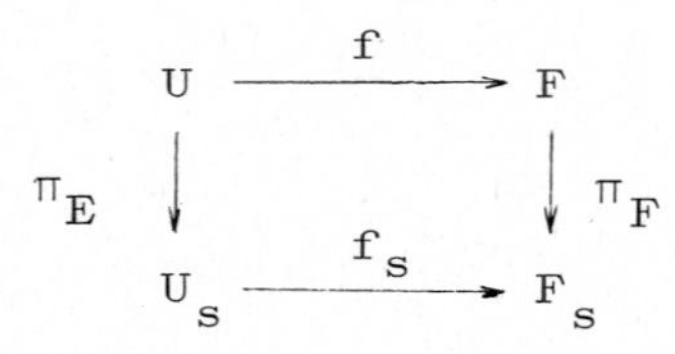

commutes. The mapping f_s is uniquely determined by f. The viewpoints of Remarks 19.1, 19.2 and 19.3 are related in the following way: if $\xi \in U$, then $d^m f(\xi)$ and the two ways of representing $d^m f_s(\xi)$ correspond under the canonical isomorphisms between the vector spaces $\mathcal{L}_s(^m E;F)/K\mathcal{L}_s(^m E;F)$, $\mathcal{L}_s(^m E;F_s)$ and $\mathcal{L}_s(^m E_s;F_s)$. A similar remark applies to $\hat{d}^m f(\xi)$ and $\hat{d}^m f_s(\xi)$.

These observations show that the study of holomorphic mappings can be reduced to the case in which E, or F, or both, are separated.

CHAPTER 20

$\mathcal{H}$-HOLOMORPHY AND H-HOLOMORPHY

We consider another way of defining a holomorphic mapping, slightly different to Definition 18.1, but not in any essential way.

DEFINITION 20.1 We define the vector space $H(U;F)$ by

$$H(U;F) = \mathcal{H}(U;\hat{F}) \cap F^U,$$

where $\hat{F}$ is a completion of F and F^U is the vector space of all mappings of U into F. Thus $H(U;F)$ consists of all mappings $f: U \to F$ which are holomorphic when we consider f as taking its values in $\hat{F}$. Clearly, $H(U;F)$ is independent of the particular choice of completion $\hat{F}$. However, assuming that F is separated, $d^m f(\xi)$ and $\hat{d}^m f(\xi)$ depend on the choice of $\hat{F}$, where $\xi \in U$, $m \in \mathbb{N}$. We have

$$\mathcal{H}(U;F) \subset H(U;F) \subset \mathcal{H}(U;\hat{F}).$$

If $f \in \mathcal{H}(U;F)$ or $f \in H(U;F)$, we say that f is $\mathcal{H}$-holomorphic or H-holomorphic respectively.

REMARK 20.1: Definition 7.1 expressses the concept of H-holomorphy in terms of $\mathcal{H}$-holomorphy. It is also possible to express $\mathcal{H}$-holomorphy in terms of H-holomorphy. Suppose, for simplicity, that F is separated - this requirement is not essential. Then $\mathcal{H}(U;F)$ consists of all $f \in H(U;F)$ such that $d^m f(\xi)(x_1,\ldots,x_m) \in F$ for every $m \in \mathbb{N}^*$, $\xi \in U$ and $x_1,\ldots,x_m \in E$, or equivalently, such that $\hat{d}^m f(\xi)(x) \in F$ for every $m \in \mathbb{N}^*$, $\xi \in U$ and $x \in E$. In the case of $\mathcal{H}$-holomorphy, we demand that f, and all its differentials at every point of U, take their values in F, whereas for H-holomorphy, f takes its values in F, while its differentials at each point may take their values in $\hat{F}$.

Whether one uses the notion of $\mathcal{H}$-holomorphy or of H-holomorphy is often simply a matter of preference or convenience; the theory of holomorphy can be developed using either notion. From the intuitive point of view, $\mathcal{H}$-holomorphy seems more natural, but from a technical viewpoint, H-holomorphy is preferable. For example, the differentials of a holomorphic mapping may be represented by integrals such as the Cauchy integral, or by infinite series, whose values, in principle, will lie in a completion, $\hat{F}$, of F, and may not belong to F, even when the mapping itself takes its values in F.

If F is complete in some suitable sense, the notions of $\mathcal{H}$-holomorphy and H-holomorphy coincide. We shall continue to work with $\mathcal{H}$-holomorphy, while mentioning H-holomorphy from time to time.

CHAPTER 21

ENTIRE MAPPINGS

DEFINITION 21.1 An element of $\mathcal{H}(E;F)$ is called an entire mapping. Thus, $f: E \to F$ is entire if f is holomorphic in E.

The following proposition shows that every continuous polynomial is entire.

PROPOSITION 21.1 $\mathcal{P}(E;F) \subset \mathcal{H}(E;F)$.

PROOF: It suffices to show that $\mathcal{P}(^kE;F) \subset \mathcal{H}(E;F)$ for every $k \in \mathbb{N}^*$. Accordingly, let $A \in \mathcal{L}_s(^kE;F)$. By Newton's formula (Lemma 14.1), we have

$$P(x) = Ax^k = A[\xi + (x-\xi)]^k = \sum_{m=0}^{k} \binom{k}{m} A\xi^{k-m}(x-\xi)^m$$

for every $x \in E$ and $\xi \in E$. It follows immediately that $P = \hat{A}$ is holomorphic in E. Q.E.D.

REMARK 21.1: If F is separated, $A \in \mathcal{L}_s(^kE;F)$, $k \in \mathbb{N}$ and $P = \hat{A}$, the proof of Proposition 21.1 shows that

$$\frac{1}{m!} d^mP(\xi) = \binom{k}{m} A\xi^{k-m} \quad \text{if} \quad m = 0,1,\ldots,k$$

and $d^mP(\xi) = 0$ if $m > k$.

If $P \in \mathcal{P}(E;F)$ has degree m, then $d^kP = 0$ if $k > m$.

LEMMA 21.1 Let F be separated. If $P \in \mathcal{P}({}^kE;F)$, $k \in \mathbb{N}$, then

$$d^m P \in \mathcal{P}({}^{k-m}E; \mathcal{L}_s({}^mE;F)) \quad \text{and}$$

$$\hat{d}^m P \in \mathcal{P}({}^{k-m}E; \mathcal{P}({}^mE;F))$$

for $m = 0,1,\ldots,k$, where $\mathcal{L}_s({}^mE;F)$ and $\mathcal{P}({}^mE;F)$ carry their limit topologies.

PROOF: Let $A \in \mathcal{L}_s({}^kE;F)$ with $P = \hat{A}$. For $m = 0,1,\ldots,k$, let $T: E^{k-m} \to \mathcal{L}_s({}^mE;F)$ be defined as follows: if $(x_1,\ldots,x_{k-m}) \in E^{k-m}$, $T(x_1,\ldots,x_{k-m})$ is the mapping which associates with $(y_1,\ldots,y_m) \in E^m$ the element $\binom{k}{m}A(x_1,\ldots,x_{k-m},y_1,\ldots,y_m)$ of F. It is easy to see that $T \in \mathcal{L}_a({}^{k-m}E; \mathcal{L}_s({}^mE;F))$, and that if $x = x_1 = \ldots = x_{k-m}$, then $T(x,x,\ldots,x) = \binom{k}{m}Ax^{k-m}$. Therefore $d^mP \in \mathcal{P}_a({}^{k-m}E; \mathcal{L}_s({}^mE;F))$.

Similarly, let $G: E^{k-m} \to \mathcal{P}({}^mE;F)$ be defined as follows: for $(x_1,\ldots,x_{k-m}) \in E^{k-m}$, $G(x_1,\ldots,x_{k-m})$ is the mapping which associates with $y \in E$ the element $\binom{k}{m}A(x_1,\ldots,x_{k-m},y,\ldots,y)$ of F. It is easy to see that $G \in \mathcal{L}_a({}^{k-m}E, \mathcal{P}({}^mE;F))$, and that if $x = x_1 = \ldots = x_{k-m}$, then $G(x,x,\ldots,x) = \widehat{\binom{k}{m}Ax^{k-m}}$. Therefore $\hat{d}^mP \in \mathcal{P}_a({}^{k-m}E; \mathcal{P}({}^mE;F))$.

Now for every $\beta \in SC(F)$ there exists $\alpha \in CS(E)$ such that $\|A\|_{\alpha,\beta} \leq 1$. Therefore, for $m \geq 1$,

$$\beta[Ax^{k-m}(x_1,\ldots,x_m)] \leq \alpha(x)^{k-m}\,\alpha(x_1)\ldots\alpha(x_m)$$

for every $x,x_1,\ldots,x_m \in E$. Hence $Ax^{k-m} \in \mathcal{L}_s({}^mE_\alpha;F_\beta)$, and

$$\|Ax^{k-m}\|_{\alpha,\beta} \leq \alpha(x)^{k-m}$$

for every $x \in E$. This also holds for $m = 0$. Therefore the $(k-m)$-homogeneous polynomial

$$x \in E_\alpha \longmapsto Ax^{k-m} \in \mathcal{L}_s({}^m E_\alpha; F_\beta)$$

is continuous, where $\mathcal{L}_s({}^m E_\alpha; F_\beta)$ carries the seminorm $\| \ \|_{\alpha,\beta}$. Since the mapping $x \in E \longmapsto x \in E_\alpha$ is continuous, the $(k-m)$-homogeneous polynomial $x \in E \longmapsto Ax^{k-m} \in \mathcal{L}_s({}^m E_\alpha; F_\beta)$ is continuous. Since the inclusion mapping of $\mathcal{L}_s({}^m E_\alpha; F_\beta)$ into $\mathcal{L}_s({}^m E; F_\beta)$ is continuous, where the second space carries the inductive topology, $x \in E \longmapsto Ax^{k-m} \in \mathcal{L}_s({}^m E; F_\beta)$ is continuous. Since this is true for every $\beta \in CS(F)$, it follows that the $(k-m)$-homogeneous polynomial $x \in E \longmapsto Ax^{k-m} \in \mathcal{L}_s({}^m E; F)$ is continuous for the limit topology on $\mathcal{L}_s({}^m E; F)$. Therefore $d^m P$ is continuous, and similarly $\hat{d}^m P$ is continuous.

Q.E.D.

REMARK 21.3: This lemma is also valid for the bounded, compact and finite topologies on $\mathcal{L}_s({}^m E; F)$ and $\mathcal{P}({}^m E; F)$, since each of these topologies is weaker than the limit topology.

CHAPTER 22

SOME ELEMENTARY PROPERTIES OF HOLOMORPHIC MAPPINGS

PROPOSITION 22.1 $\mathcal{H}(U;F)$ increases if the topology of E is strengthened and the topology of F weakened; if the topology of F remains separated, the differentials of a holomorphic mapping of U into F are unchanged.

PROOF: Let $\tau_j(E)$ and $\tau_j(F)$, $j = 1,2$, be locally convex topologies on the complex vector spaces E and F respectively. Let E_j and F_j, $j = 1,2$, respectively denote the spaces E and F with the topologies $\tau_j(E)$ and $\tau_j(F)$. Suppose that $\tau_1(E) \leq \tau_2(E)$ and $\tau_1(F) \geq \tau_2(F)$. If U is a non-empty open subset of E_1, then U is also open in E_2. Let U_1 and U_2 denote U considered as an open subset of E_1, and of E_2, respectively. Then clearly $\mathcal{H}(U_1;F_1) \subset$ $\subset \mathcal{H}(U_2;F_2)$. Furthermore, if $f \in \mathcal{H}(U_1;F_1)$ and F_2 is separated, it follows from the uniqueness of the Taylor series that the differentials $d^m f$ and $\hat{d}^m f$ are the same whether we consider f as a mapping of U into F_1, or as a mapping of U into F_2. Q.E.D.

REMARK 22.1: It is easy to see that the first part of this proposition is also true for H-holomorphic mappings: The second part takes the following form: Let $\hat{F}_1$ and $\hat{F}_2$ be completions of F_1 and F_2 respectively. Then the continuous identity mapping of F_1 into F_2 extends to a continuous linear mapping $\pi: \hat{F}_1 \to \hat{F}_2$. Then if $f \in H(U;F_1)$, and $d^m f(\xi)$, $\hat{d}^m f(\xi)$ are its differentials at $\xi \in U$, then $\pi \circ d^m f(\xi)$ and $\pi \circ \hat{d}^m f(\xi)$ are the differentials of f at ξ, when f is considered as an element of $H(U;F_2)$.

Proposition 22.1 is a special case of Propositions 22.2 and 22.3. These are themselves special cases of propositions concerning composition of holomorphic mappings.

PROPOSITION 22.2 Let E, F and G be complex locally convex spaces, let $f \in \mathcal{H}(U;F)$ and let $g \in \mathcal{H}(F;G)$ be a continuous affine mapping; that is, $g(x) = A + B(x)$, where $A \in G$ and $B \in \mathcal{L}(F;G)$. Then $g \circ f \in \mathcal{H}(U;G)$, and if F and G are separated,

$$d^m(g \circ f)(\xi) = B \circ d^m f(\xi) \quad \text{and}$$

$$\hat{d}^m(g \circ f)(\xi) = B \circ \hat{d}^m f(\xi) \quad \text{for } m \in \mathbb{N},\ \xi \in U.$$

PROOF: The proof follows easily from the definition of a holomorphic mapping and its differentials. Q.E.D.

PROPOSITION 22.3 Let E, F and G be complex locally convex spaces, and let V be a non-empty open subset of F. If $f \in \mathcal{H}(E;F)$ is a continuous affine mapping, so that $f(x) = A + B(x)$, where $A \in F$ and $B \in \mathcal{L}(E;F)$, if $g \in \mathcal{H}(V;G)$, and if $U = f^{-1}(V)$ is non-empty, then $g \circ f \in \mathcal{H}(U;G)$. If F and G are separated, then

$$d^m(g \circ f)(\xi) = d^m g[f(\xi)] \circ B^m \quad \text{and}$$

$$\hat{d}^m(g \circ f)(\xi) = \hat{d}^m g[f(\xi)] \circ B^m \quad \text{for } m \in \mathbb{N},\ \xi \in U,$$

where $B^m: E^m \to F^m$ is defined by

$$B^m(x_1,\ldots,x_m) = (B(x_1),\ldots,B(x_m)), \quad x_1,\ldots,x_m \in E.$$

PROOF: The proof follows easily from the definition of a holomorphic mapping and its differentials. Q.E.D.

The proofs of the following propositions are immediate.

PROPOSITION 22.4 Let E and F_λ, $\lambda \in \Lambda$, $\Lambda \neq \emptyset$, be complex locally convex spaces, and let $F = \prod_{\lambda \in \Lambda} F_\lambda$. If U is a nonempty open subset of E, and $f: U \to F$ has as its components the mappings $f_\lambda : U \to F_\lambda$, then $f \in \mathcal{H}(U;F)$ if and only if $f_\lambda \in \mathcal{H}(U;F_\lambda)$ for every $\lambda \in \Lambda$. Furtheremore, if F_λ is separated for every $\lambda \in \Lambda$, then for every $\xi \in U$, $d^m f(\xi)$ corresponds to $(d^m f_\lambda(\xi))_{\lambda \in \Lambda}$ under the canonical isomorphism of $\mathcal{L}_s(^m E;F)$ with $\prod_{\lambda \in \Lambda} \mathcal{L}_s(^m E;F_\lambda)$. Similarly, $\hat{d}^m f(\xi)$ corresponds to $(\hat{d}^m f_\lambda(\xi))_{\lambda \in \Lambda}$.

PROPOSITION 22.5 Let E be a complex locally convex space, and F a complex vector space. Let $\{\tau_\lambda(F)\}_{\lambda \in \Lambda}$, $\Lambda \neq \emptyset$, be a family of locally convex topologies on F, and let F_λ denote the locally convex space $(F,\tau_\lambda(F))$. If τ denotes the locally convex topology $\sup_{\lambda \in \Lambda} \tau_\lambda(F)$ on F, then

$$\mathcal{H}(U;(F,\tau)) = \bigcap_{\lambda \in \Lambda} \mathcal{H}(U;F_\lambda).$$

CHAPTER 23

HOLOMORPHY, CONTINUITY AND AMPLE BOUNDEDNESS

PROPOSITION 23.1 $\mathcal{H}(U;F) \subset H(U;F) \subset C(U;F)$, where $C(U;F)$ is the vector space of continuous mappings of U into F.

PROOF: We show first that $\mathcal{H}(U;F) \subset C(U;F)$. Let $f \in \mathcal{H}(U;F)$, and let $\xi \in U$. Consider the Taylor polynomials

$$s_m(x) = \sum_{k=0}^{m} A_k(x-\xi)^k$$

of f at ξ. Since f is holomorphic, for every $\beta \in CS(F)$ there exists a neighbourhood V of ξ in U such that

$$(*) \qquad \lim_{m\to\infty} \beta[f(x)-s_m(x)] = 0 \quad \text{uniformly in} \quad x \in V.$$

Since $s_m \in C(E;F)$, it follows from (*) that $f/V \in C(V;F_\beta)$ where V has the induced topology from U. Therefore $f \in C(U;F_\beta)$ for every $\beta \in CS(F)$, and so $f \in C(U;F)$.

To see that $H(U;F) \subset C(U;F)$, let $\hat{F}$ be a completion of F. Then, from what we have already proved, $H(U;F) \subset \mathcal{H}(U;\hat{F}) \subset C(U;\hat{F})$. Since $H(U;F) \subset F^U$, it follows that $H(U;F) \subset C(U;F)$. Q.E.D.

REMARK 23.1: The assumption of continuity of the Taylor coefficients in Definition 18.1 implies that every holomorphic mapping is continuous. The next proposition gives a very useful characterization of holomorphy.

PROPOSITION 23.2 In order that $f: U \to F$ be holomorphic, it is necessary and sufficient that f be either continuous or amply bounded, and that, for every $\xi \in U$ there exists a sequence $A_m \in \mathcal{L}_{as}(^mE;F)$, $m \in \mathbb{N}$, with the following property: for every $\beta \in CS(F)$ there exists a neighbourhood V of ξ in U such that

$$\lim_{m\to\infty} \beta[f(x) - \sum_{k=0}^{m} A_k(x-\xi)^k] = 0$$

uniformly in $x \in V$.

PROOF: Necessity follows from Definition 18.1 and Proposition 23.1 and the fact that continuity implies ample boundedness.

To prove that the stated condition is sufficient, suppose that $f: U \to F$ is amply bounded, and that for each $\xi \in U$ there is a sequence $A_m \in \mathcal{L}_{as}(^mE;F)$ with the given property. We must show that A_m is continuous, for every $m \in \mathbb{N}$. Let $s_m(x) = \sum_{k=0}^{m} A_k(x-\xi)^k$, $x \in E$. Then $s_m \in \mathcal{P}_a(E;F)$. For every $\beta \in CS(F)$ let V be a neighbourhood of ξ in U such that $\lim_{m\to\infty} \beta[f(x)-s_m(x)] = 0$ uniformly in $x \in V$ and $\sup_{x\in V} \beta[f(x)]<\infty$. Then there exists $C > 0$ such that $\beta[f(x)-s_m(x)] \leq C$ for every $m \in \mathbb{N}$, $x \in V$. It follows that $\sup_{x\in V} \beta[s_m(x)] < \infty$ for every $m \in \mathbb{N}$. Thus s_m is amply bounded at ξ, and so, by Proposition 15.3, s_m is continuous for every $m \in \mathbb{N}$. Hence $A_m = s_m - s_{m-1}$ is continuous for every $m \in \mathbb{N}$.

Finally, if f is continuous, then f is amply bounded, and so the stated condition is sufficient. Q.E.D.

CHAPTER 24

BOUNDING SETS

REMARK 24.1: Let E'_a denote the algebraic dual of E. If $\varphi \in E'_a$, $b \in F$ and $m \in N$, then $\varphi^m \cdot b \in \mathcal{P}_a(^mE;F)$ is defined by

$$[\varphi^m \cdot b](x) = [\varphi(x)]^m b \quad \text{for} \quad x \in E.$$

When φ belongs to E', the topological dual of E, we have $\varphi^m \cdot b \in \mathcal{P}(^mE;F)$. For $m = 0$, we define $\varphi^o \cdot b$ to be the constant mapping $x \in E \longmapsto b \in F$. $\varphi^m \cdot b$ is the simplest possible example of an m-homogeneous polynomial.

Given $\varphi_m \in E'$ and $b_m \in F$ $(m \in \mathbb{N}^*)$, and assuming that F is separated, we seek conditions under which the series

$$(1) \qquad \sum_{m=1}^{\infty} [\varphi_m(x)]^m b_m$$

converges in F for every $x \in E$ and defines a holomorphic mapping, f, of E into F, where

$$(2) \qquad f(x) = \sum_{m=1}^{\infty} [\varphi_m(x)]^m b_m .$$

If this is so, then (1) is the Taylor series of f at 0.

Consider the set I of all $m \in \mathbb{N}^*$ for which $b_m \neq 0$. If I is finite, then $f \in \mathcal{P}(E);F) \subset \mathcal{H}(E;F)$. Suppose that I is infinite. If F is normed we may assume that $\|b_m\| = 1$ for every $m \in I$, since φ_m and b_m can be replaced by $\|b_m\|^{1/m}\varphi_m$ and $b_m/\|b_m\|$ respectively. This motivates the condition on the sequence $\{b_m\}_{m\in\mathbb{N}^*}$ in the next proposition. We remark that in the case of a general separated space F, even when F is metrizable it is not necessarily true that, given $b_m \in F$, $b_m \neq 0$ for $m \in I$ where $I \subset \mathbb{N}^*$ is infinite, there is a sequence $\lambda_m \in C$, $m \in I$, such that $\{\lambda_m b_m\}_{m\in I}$ is bounded and bounded away from zero. We recall that a sequence $\{a_m\}_{m\in\mathbb{N}}$ in a separated space F is said to be bounded away from zero if there exist $\alpha \in CS(F)$ and $\delta > 0$ such that $\alpha(a_m) \geq \delta$ for every $m \in \mathbb{N}$.

PROPOSITION 24.1 Let F be separated and sequentially complete, let $\varphi_m \in E'$, and $b_m \in F$ for $m \in \mathbb{N}$, and let $I = \{m \in \mathbb{N}^* : b_m \neq 0\}$ be infinite. Suppose that $\{b_m\}_{m\in I}$ is bounded and bounded away from zero. Then in order that the series

$$1) \qquad \sum_{m\in I} [\varphi_m(x)]^m \cdot b_m$$

converges for every $x \in E$, and defines an entire mapping, it is necessary and sufficient that $\varphi_m(x) \to 0$ as $m \to \infty$ for every $x \in E$, and that the sequence $\{\varphi_m\}_{m\in I}$ is equicontinuous, that is, that φ_m converges to zero in the weak topology $\sigma(E',E)$, and is equicontinuous.

PROOF: Sufficiency: For each $x \in E$ there exists $M_x \in I$ such that $|\varphi_m(x)| \leq 1/2$ for $m \geq M_x$. If $\beta \in CS(F)$, there exists $C \geq 0$ such that $\beta(b_m) \leq C$ for every m. Hence

$\beta\{[\varphi_m(x)]^m \cdot b_m\} = |\varphi_m(x)|^m \cdot \beta(b_m) \le C2^{-m}$ for $m \ge M_x$. Therefore the series 1) converges for every $x \in E$. We must show that $f(x) = \sum_{m \in I} [\varphi_m(x)]^m \cdot b_m$ defines an entire mapping of E into F. Let $\xi \in E$. There exists $M \in I$ such that $|\varphi_m(\xi)| \le 1/3$ for $m \ge M$, and so there exists $D \ge 1$ such that $|\varphi_m(\xi)|^m \le D(1/3)^m$ for every $m \in \mathbb{N}^*$. Since $\{\varphi_m\}_{m \in \mathbb{N}^*}$ is equicontinuous, there exists a neighbourhood V of the origin in E such that if $t \in V$ then $|\varphi_m(t)| \le 1/3$ for every $m \in \mathbb{N}^*$. We claim that the family

$$2) \quad \{\binom{m}{h} [\varphi_m(\xi)]^{m-h}[\varphi_m(t)]^h \cdot b_m\} \quad m \in \mathbb{N}^*, \quad 0 \le h \le m$$

is uniformly absolutely summable in V. To show this, we prove that for every $\beta \in CS(F)$, the value of β on the sum of any finite subset of the family 2) is uniformly bounded for $t \in V$. Given such a finite subset, let m_o be the greatest value of m which occurs. Then the value of β on the sum of the elements of this finite subset is at most

$$\sum_{0 \le h \le m \le m_o} \beta\{\binom{m}{h} [\varphi_m(\xi)]^{m-h}[\varphi_m(t)]^h \cdot b_m\}$$

$$\le C \sum_{0 \le h \le m \le m_o} \binom{m}{h} |\varphi_m(\xi)|^{m-h} |\varphi_m(t)|^h$$

$$= C \sum_{0 \le m \le m_o} \{|\varphi_m(\xi)| + |\varphi_m(t)|\}^m$$

$$\le C \sum_{0 \le m \le m_o} \{\frac{D^{1/m}}{3} + \frac{1}{3}\}^m \le CD \sum_{0 \le m \le m_o} \{1/3 + 1/3\}^m$$

$$\le 3CD,$$

for every $t \in V$, where $\beta(b_m) \le C$ for all $m \in \mathbb{N}^*$. Therefore the family 2) is uniformly absolutely summable in V, and hence

$$\sum_{m\in I}\ \sum_{0\leq h\leq m} \binom{m}{h} [\varphi_m(\xi)]^{m-h}[\varphi_m(t)]^h b_m =$$

$$\sum_{h\in\mathbb{N}}\ \sum_{\substack{m\in I\\ m\geq h}} \binom{m}{h} [\varphi_m(\xi)]^{m-h}[\varphi_m(t)]^h b_m .$$

That is, $f(\xi+t) = \sum_{h\in\mathbb{N}} P_h(t)$ uniformly in V, where

$$P_h(t) = \sum_{\substack{m\in I\\ m\geq h}} \binom{m}{h} [\varphi_m(\xi)]^{m-h}[\varphi_m(t)]^h b_m .$$

It is clear from what we have shown that this series converges. We shall prove that P_h is a continuous h-homogeneous polynomial for every h, from which it follows that f is holomorphic at ξ. Consider the mapping

$$(t_1,\ldots,t_h) \in E^h \mapsto \sum_{\substack{m\in I\\ m\geq h}} \binom{m}{h}[\varphi_m(\xi)]^{m-h}\varphi_m(t_1)\ldots\varphi_m(t_h)b_m \in F;$$

this series is absolutely summable in F, and defines a symmetric h-linear mapping from E^h into F whose associated polynomial is P_h. And if $t_o \in E$, there exists a neighbourhood V of the origin in E such that for $t \in t_o + V$, we have $|\varphi_m(t)| \leq 1/3$ for every $m \in \mathbb{N}^*$. Thus the series defining P_h converges uniformly in a neighbourhood of t_o, and so P_h is continuous.

Necessity: Suppose that $f \in \mathcal{H}(E;F)$. Then by the uniqueness of the Taylor series, $\sum_{m\in I} (\varphi_m)^m b_m$ is the Taylor series of f at zero. Since the sequence $\{b_m\}_{m\in I}$ is bounded away from zero, there exists $\beta \in CS(F)$ such that

$\beta(b_m) \geq 1$ for $m \in I$. There is a neighbourhood V of the origin in E such that

$$\lim_{m\to\infty} \beta\{f(x) - \sum_{k\in I,k=0}^{m} [\varphi_k(x)]^k b_k = 0$$

uniformly for $x \in V$. It follows, by the Cauchy condition for convergence, that

$$\lim_{m\to\infty} \beta[\varphi_m(x)]^m b_m\} = 0$$

uniformly for $x \in V$, and hence

$$\lim_{m\to\infty} |\varphi_m(x)|^m = 0$$

uniformly for $x \in V$. Therefore there exists $M \in \mathbb{N}^*$ such that $|\varphi_m(x)|^m \leq 1$ for $m \geq M$ and $x \in V$, that is, $|\varphi_m(x)| \leq 1$ for $m \geq M$, $x \in V$. Therefore $\{\varphi_m\}_{m\in\mathbb{N}^*}$ is equicontinuous. Furthermore, since the series $\sum_{m\in I} [\varphi_m(x)]^m b_m$ converges for every $x \in E$, $[\varphi_m(x)]^m b_m \to 0$ as $m \to \infty$, for every $x \in E$. With $\beta \in CS(F)$ as above, we then have $\beta\{[\varphi_m(x)]^m b_m\} \to 0$ as $m \to \infty$, for every $x \in E$, and hence $[\varphi_m(x)]^m \to 0$ as $m \to \infty$, for every $x \in E$. Replacing x by λx, we find that

$$[\varphi_m(x)]^m \lambda^m \to 0 \quad \text{as} \quad m \to \infty$$

for every $x \in E$, and $\lambda \in \mathbb{C}$. Therefore $\varphi_m(x) \to 0$ as $m \to \infty$ for every $x \in E$. Q.E.D.

REMARK 24.2: If the space E is barrelled, then the condition $\varphi_m \to 0$ in the weak topology $\sigma(E',E)$ implies that $\{\varphi_m\}_{m\in\mathbb{N}}$ is equicontinuous.

DEFINITION 24.1 A subset X of E is said to be $\mathcal{H}$-bounding in E if every $f \in \mathcal{H}(E) = \mathcal{H}(E;\mathbb{C})$ is bounded on X. For example, a compact subset of E is $\mathcal{H}$-bounding in E. A subset of a $\mathcal{H}$-bounding set is $\mathcal{H}$-bounding, and the closure of a $\mathcal{H}$-bounding set is $\mathcal{H}$-bounding. Since $E' \subset \mathcal{H}(E)$, every $\mathcal{H}$-bounding subset of E is bounded.

REMARK 24.3: If $X \subset E$ is $\mathcal{H}$-bounding in E, then for every space F, every $f \in \mathcal{H}(E;F)$ is bounded on X. This follows from the fact that for every $\psi \in F'$, $\psi \circ f \in \mathcal{H}(E)$ is bounded on X; hence $f(X)$ is bounded in the weak topology $\sigma(F,F')$, and therefore $f(X)$ is bounded in the original topology of F. Conversely, if every $f \in \mathcal{H}(E;F)$ is bounded on X for some separated space $F \neq \{0\}$, then X is $\mathcal{H}$-bounding in E. To see this, choose $b \in F$, $b \neq 0$. Then for every $f \in \mathcal{H}(E)$, $f \cdot b \in \mathcal{H}(E;F)$ is bounded on X. Thus $f(X) \cdot b$ is bounded in F, and it follows that $f(X)$ is bounded in $\mathbb{C}$.

REMARK 24.4: If E is a semi-Montel space, so that every bounded subset of E is relatively compact, then the subsets of E which are $\mathcal{H}$-bounding in E are precisely the bounded sets.

LEMMA 24.1 Suppose that the family $\{a_{m,n}\}_{m,n\in\mathbb{N}}$ of complex numbers is unbounded, and that

$$\sum_{m\in\mathbb{N}} |a_{m,n}| < \infty \quad \text{for every} \quad n \in \mathbb{N}.$$

Then there exist sequences $m_o < m_1 < \ldots < m_h < \ldots$ and $n_o < n_1 < \ldots < n_k < \ldots$ in $\mathbb{N}$ such that

$$\sum_{h\in\mathbb{N}} a_{m_h,n_k} \to \infty \quad \text{as} \quad k \to \infty.$$

PROPOSITION 24.2 Let X be $\mathcal{H}$-bounding in E. Then for every equicontinuous sequence $\{\varphi_m\}_{m\in\mathbb{N}}$ in E' which converges to 0 in the weak topology $\sigma(E',E)$,

$$\sup\{|\varphi_m(x)| : x \in X\} \to 0 \quad \text{as} \quad m \to \infty.$$

PROOF: Suppose the conclusion is false. Then there is an equicontinuous sequence $\{\varphi_m\}_{m\in\mathbb{N}}$ in E' which converges to 0 in the weak topology $\sigma(E',E)$, such that

$$\sup\{|\varphi_m(x)| : x \in X\} \not\to 0 \quad \text{as} \quad m \to \infty,$$

where X is $\mathcal{H}$-bounding in E. Passing to a subsequence if necessary, we may assume that there exists $\delta > 0$ such that for every $m \in \mathbb{N}$ there exists $x_m \in X$ for which

$$|\varphi_m(x_m)| \geq \delta.$$

Let $a_{m,n} = [\phi_m(x_n)]^m$, where $\phi_m = \frac{2\varphi_m}{\delta}$, $m,n \in \mathbb{N}$. It is easy to see that $\{a_{m,n}\}_{m,n\in\mathbb{N}}$ is unbounded; indeed, $\{a_{m,n}\}_{m\in\mathbb{N}}$ is unbounded for every n. Furthermore, since $\varphi_m(x_n) \to 0$ as $m \to \infty$ for every n, we have

$$\sum_{m\in\mathbb{N}} |a_{m,n}| < \infty \quad \text{for every} \quad n \in \mathbb{N}.$$

Applying Lemma 24.1, there exist $m_o < m_1 < \ldots < m_h < \ldots$ and $n_o < n_1 < \ldots < n_k < \ldots$ in $\mathbb{N}$ such that

$$\sum_{h\in\mathbb{N}} [\phi_{m_h}(x_{n_k})]^{m_h} \to \infty \quad \text{as} \quad k \to \infty.$$

Let $f = \sum_{h\in\mathbb{N}} [\phi_{m_h}]^{m_h}$. By Proposition 24.1 we have $f \in \mathcal{H}(E)$, but f is unbounded on $\{x_{n_k}\}_{k\in\mathbb{N}}$ and hence is unbounded on X.

Q.E.D.

COROLLARY 24.1 If E' contains an equicontinuous sequence $\{\varphi_m\}_{m\in\mathbb{N}}$ such that φ_m converges to 0 in the weak topology $\sigma(E',E)$ but does not converge to 0 in the strong topology $\beta(E',E)$, then E contains a bounded set which is not $\mathcal{H}$-bounding.

REMARK 24.5: Josefson and Nissenzweig, in 1973, [63], [112] proved the following theorem which provided the solution to an important and long-standing problem in the theory of normed spaces.

PROPOSITION 24.3 If E is a normed space of infinite dimension, there exists a sequence $\{\varphi_m\}_{m\in\mathbb{N}}$ in E' such that $\|\varphi_m\| = 1$ for every $m \in \mathbb{N}$ and $\varphi_m \to 0$ as $m \to \infty$ in the weak topology $\sigma(E',E)$.

This result enables us to prove the following proposition.

PROPOSITION 24.4 Let E be a complex normed space of infinite dimension. Then every $\mathcal{H}$-bounding set in E has empty interior.

PROOF: By Proposition 24.3, there is a sequence $\{\varphi_m\}_{m\in\mathbb{N}}$ in E which converges weakly to zero, and $\|\varphi_m\| = 1$ for every $m \in \mathbb{N}$. Let $\varphi: E \to \mathbb{C}$ be the mapping

$$\varphi(x) = \sum_{m\in\mathbb{N}^*} [\varphi_m(x)]^m.$$

By Proposition 24.1, $\varphi \in \mathcal{H}(E)$. We claim that φ is unbounded on $\bar{B}_r(0)$, for $r > 1$. Suppose this were not so. Then

$$M = \sup\{|\varphi(x)| : \|x\| \leq r\} < \infty$$

for some $r > 1$. Using the Cauchy inequalities we have

$$\left\|\frac{1}{m!}\hat{d}^m\varphi(0)\right\| \leq \frac{1}{r^m} \sup_{\|x\|\leq r} \|\varphi(x)\| \leq \frac{M}{r^m}.$$

Therefore $\|\varphi_m\|^m \leq M/r^m$ for every $m \in \mathbb{N}^*$; hence $\|\varphi_m\| \leq$ $\leq \frac{M^{1/m}}{r}$ for every $m \in \mathbb{N}^*$, but this implies that $1 < r \leq M^{1/m}$ for every $m \in \mathbb{N}^*$, which is absurd. Thus we have shown that for every closed ball $\bar{B}_r(0)$ with $r > 1$, there exists $f \in \mathcal{H}(E)$ which is unbounded on $\bar{B}_r(0)$. It follows easily that the same is true of the closed balls $\bar{B}_{r_1}(0)$ where $r_1 > 0$, and hence, by translation, it follows that for every $\xi \in E$ and every $r > 0$ there exists $f \in \mathcal{H}(E)$ which is unbounded on $\bar{B}_r(\xi)$. Q.E.D.

LEMMA 24.2 Let E be a complex locally convex space, and let X be a subset of E which is bounded and not precompact. Then there exist a sequence $\{x_m\}_{m\in\mathbb{N}}$ in X, $\alpha \in CS(E)$, and $\delta > 0$ such that, if S_m is the subspace of E generated by $x_o,\ldots,x_m$, then $\alpha(u-x_{m+1}) \geq \delta$ for every $u \in S_m$, $m \in \mathbb{N}$.

PROOF: Since X is not precompact, there is a sequence $\{t_p\}_{p\in\mathbb{N}}$ in X, $\alpha \in CS(E)$ and $\gamma > 0$, such that

1) $\alpha(t_p-t_q) \geq \gamma$ for $p,q \in \mathbb{N}$, $p \neq q$.

Fix δ, $0 < \delta < \gamma/2$. We claim that if S is a finite dimensional subspace of E, there exists $p \in \mathbb{N}$ such that

$\alpha(u-t_p) \geq \delta$ for every $u \in S$. Suppose this were not so; then for every $p \in \mathbb{N}$, there exists $u_p \in S$ such that $\alpha(u_p-t_p) < \delta$. Since $\{\alpha(t_p)\}_{p\in\mathbb{N}}$ is bounded, $\{\alpha(u_p)\}_{p\in\mathbb{N}}$ is also bounded, and so, since S is finite dimensional, there exist $p_o < p_1 < \ldots < p_h < \ldots$ in $\mathbb{N}$ and $u \in S$ such that $\alpha(u-u_{p_h}) \to 0$ as $h \to \infty$. Then

$$\alpha(u-t_{p_h}) \leq \alpha(u-u_{p_h}) + \alpha(u_{p_h}-t_{p_h}) < (\gamma/2-\delta) + \delta < \gamma/2$$

for h sufficiently large, which contradicts 1). Thus our claim is established.

We now define the sequence $\{x_m\}_{m\in\mathbb{N}}$ recursively. Let $x_o = t_o$, and suppose $x_o,\ldots,x_m$ have been chosen to satisfy the desired condition. Let S be the subspace of E generated by $x_o,\ldots,x_m$. Then there exists $p_m \in \mathbb{N}$ such that $\alpha(u-t_{p_m}) \geq \delta$ for every $u \in S$. We define $x_{m+1} = t_{p_m}$. Then the sequence $\{x_m\}_{m\in\mathbb{N}}$ has the stated property. Q.E.D.

PROPOSITION 24.5 (Dineen-Hirschowitz). If every equicontinuous sequence in E' contains a subsequence which converges in the weak topology $\sigma(E',E)$ then every $\mathcal{H}$-bounding set in E is precompact.

PROOF: Let X be a subset of E which is not precompact; we must show that X is not $\mathcal{H}$-bounding in E. Since every $\mathcal{H}$-bounding set is bounded, we may assume that X is bounded. Let $\{x_m\}_{m\in\mathbb{N}} \subset X$, $\alpha \in CS(E)$ and $\delta > 0$ be as in Lemma 24.2. By the Hahn-Banach theorem, there exist $\varphi_m \in E'$, $m \in \mathbb{N}$, such that $\varphi_m/S_m = 0$, $\varphi_m(x_{m+1}) = 1$, and $|\varphi_m(x)| \leq \frac{1}{\delta}\alpha(x)$ for every $x \in E$. Therefore the sequence $\{\varphi_m\}_{m\in\mathbb{N}}$ is equicontinuous and so, by hypothesis, there exist $m_0 < m_1 < \ldots < m_h < \ldots$ in $\mathbb{N}$ and $\varphi \in E'$ such that $\{\varphi_{m_h}\}_{h\in\mathbb{N}}$ converges to φ in $\sigma(E',E)$. Let $\psi_h = \varphi_{m_h} - \varphi$. Since $m_k \geq k$, we have $\varphi_{m_k}(x_h) = 0$ for $h \leq k$, and so $\varphi(x_h) = 0$ for every $h \in \mathbb{N}$. Therefore

$$\psi_h(x_{m_h+1}) = \varphi_{m_h}(x_{m_h+1}) - \varphi(x_{m_h+1}) = \varphi_{m_h}(x_{m_h+1}) = 1$$

for every $h \in \mathbb{N}$. Thus $\{\psi_h\}_{h\in\mathbb{N}}$ is equicontinuous, and ψ_h converges to 0 in $\sigma(E',E)$, but

$$\sup\{|\psi_h(x)| : x \in X\} \geq 1$$

for every $h \in \mathbb{N}$. Therefore, by Proposition 24.2, X is not $\mathcal{H}$-bounding in E. Q.E.D.

REMARK 24.6: In general, a precompact subset of E need not be $\mathcal{H}$-bounding in E. Clearly, if E has the property that every closed precompact subset is compact, then every precompact subset is $\mathcal{H}$-bounding in E. In particular, this is true if E is complete. We are thus led to a new concept of completeness: if E satisfies the hypothesis of Proposition 24.5 and if every closed precompact subset of E is compact, then the $\mathcal{H}$-bounding subsets of E are precisely the precompact sets.

EXAMPLE 24.1 (see Dineen). Let $E = \ell^\infty(\mathbb{N})$, and for each $m \in \mathbb{N}$, let e_m be the element of E whose m-th coordinate is 1, with every other coordinate equal to 0. Then $\{e_m\}_{m\in\mathbb{N}}$ is $\mathcal{H}$-bounding in E, but is not precompact. Evidently, E does not satisfy the conditions of Proposition 24.5.

CHAPTER 25

THE CAUCHY INTEGRAL AND THE CAUCHY INEQUALITIES

In this chapter we shall be considering integrals of functions from subsets of $\mathbb{C}$ into F. Although in principle these integrals will take their values in a completion, $\hat{F}$, of F, the equations we derive show that the integrals in question lie in F.

PROPOSITION 25.1 (The Cauchy Integral). Let F be separated, $f \in \mathcal{H}(U;F)$, $\xi \in U$, $x \in U$ and $\rho > 1$ such that $(1-\lambda)\xi + \lambda x \in U$ for every $\lambda \in \mathbb{C}$, $|\lambda| \leq \rho$. Then

$$f(x) = \frac{1}{2\pi i} \int_{|\lambda|=\rho} \frac{f[(1-\lambda)\xi+\lambda x]}{\lambda-1} d\lambda .$$

PROOF: Suppose that V is an open subset of $\mathbb{C}$, $\tau \in B_\delta(\zeta) \subset \bar{B}_\delta(\zeta) \subset V$ for some $\delta > 0$ and $g \in \mathcal{H}(V;F)$. Then

$$g(\lambda) = \frac{1}{2\pi i} \int_{|\lambda-\zeta|=\delta} \frac{g(\lambda)}{\lambda-\tau} d\lambda .$$

This can be proved in exactly the same way as the classical case $F = \mathbb{C}$, or by reducing the case of general F to that of $F = \mathbb{C}$ by means of the Hahn-Banach theorem.

Now let $g(\lambda) = f[(1-\lambda)\xi+\lambda x]$. Then g is defined, and is holomorphic, in the open subset V of $\mathbb{C}$ consisting of all $\lambda \in \mathbb{C}$ for which $(1-\lambda)\xi + \lambda x \in U$. By hypothesis, $1 \in B_\rho(0) \subset \bar{B}_\rho(0) \subset V$, and hence

$$g(1) = \frac{1}{2\pi i} \int_{|\lambda|=\rho} \frac{g(\lambda)}{\lambda - 1}\, d\lambda,$$

that is,

$$f(x) = \frac{1}{2\pi i} \int_{|\lambda|=\rho} \frac{f[(1-\lambda)\xi + \lambda x]}{\lambda - 1}\, d\lambda.$$

Q.E.D.

PROPOSITION 25.2 (The Cauchy Integral). Let F be separated, $f \in \mathcal{H}(U;F)$, $\xi \in U$, $x \in E$, $m \in \mathbb{N}$ and $\rho > 0$ such that $\xi + \lambda x \in U$ for every $\lambda \in \mathbb{C}$, $|\lambda| \le \rho$. Then

$$\frac{1}{m!} d^m f(\xi) x^m = \frac{1}{2\pi i} \int_{|\lambda|=\rho} \frac{f(\xi + \lambda x)}{\lambda^{m+1}}\, d\lambda.$$

PROOF: Suppose that $V \subset \mathbb{C}$ is open, and the closed annulus with centre ζ and radii r and R, $0 < r \le R$, is contained in V. Then, if $g \in \mathcal{H}(V;F)$,

$$\int_{|\lambda - \zeta| = r} g(\lambda)\, d\lambda = \int_{|\lambda - \zeta| = R} g(\lambda)\, d\lambda.$$

This can be proved exactly as in the classical case $F = \mathbb{C}$, or by reducing the case of general F to that of $F = \mathbb{C}$ by means of the Hahn-Banach theorem.

Now let $g(\lambda) = \dfrac{f(\xi + \lambda x)}{\lambda^{m+1}}$. Then g is defined, and is holomorphic, in the open set $V \subset \mathbb{C}$ consisting of all $\lambda \in \mathbb{C}$ such that $\xi + \lambda x \in U$ and $\lambda \ne 0$. By hypothesis, $\bar{B}_\rho(0) - \{0\} \subset \subset V$. Therefore, if $0 < \varepsilon \le \rho$, we have

$$\int_{|\lambda|=\varepsilon} g(\lambda)d\lambda = \int_{|\lambda|=\rho} g(\lambda)d\lambda ,$$

that is,

$$\int_{|\lambda|=\varepsilon} \frac{f(\xi+\lambda x)}{\lambda^{m+1}}\, d\lambda = \int_{|\lambda|=\rho} \frac{f(\xi+\lambda x)}{\lambda^{m+1}}\, d\lambda .$$

For $n \in \mathbb{N}$, $t \in E$, let

$$P_n = \frac{1}{n!}\, \hat{d}^n f(\xi) \quad \text{and} \quad s_n(t) = \sum_{k=0}^{n} P_k(t-\xi) .$$

Then for $n \geq m$,

$$\int_{|\lambda|=\rho} \frac{s_n(\xi+\lambda x)}{\lambda^{m+1}} = \sum_{k=0}^{n} P_k(x) \int_{|\lambda|=\varepsilon} \frac{d\lambda}{\lambda^{m+1-k}} = 2\pi i\, P_m(x) .$$

Therefore, for $n \geq m$, we have

$$\int_{|\lambda|=\rho} \frac{f(\xi+\lambda x)}{\lambda^{m+1}}\, d\lambda - 2\pi i\, P_m(x) = \int_{|\lambda|=\varepsilon} \frac{f(\xi+\lambda x)-s_n(\xi+\lambda x)}{\lambda^{m+1}}\, d\lambda$$

and hence, for every $\beta \in CS(F)$,

$$\beta[\int_{|\lambda|=\rho} \frac{f(\xi+\lambda x)}{\lambda^{m+1}}\, d\lambda - 2\pi i\, P_m(x) \leq$$

$$\leq \frac{1}{\varepsilon^m} \sup_{|\lambda|=\varepsilon} \beta[f(\xi+\lambda x)-s_n(\xi+\lambda x)] .$$

But for every $\beta \in CS(F)$ there is a neighbourhood U' of ξ in U such that $\lim_{n\to\infty} \beta[f(y)-s_n(y)] = 0$ uniformly on U'. Therefore, if we choose $\varepsilon > 0$ sufficiently small and let $n \to \infty$, we obtain

$$\beta[\int_{|\lambda|=\rho} \frac{f(\xi+\lambda x)}{\lambda^{m+1}}\,d\lambda - 2\pi i\, P_m(x)] = 0.$$

Since this is true for every $\beta \in CS(F)$, we have

$$\int_{|\lambda|=\rho} \frac{f(\xi+\lambda x)}{\lambda^{m+1}}\,d\lambda = 2\pi i\, P_m(x).$$

Q.E.D.

REMARK 25.1: The case $m = 0$ of Proposition 25.2 is equivalent to Proposition 25.1; this can be seen by performing a change of variable.

REMARK 25.2: If $\alpha \in CS(E)$, $\beta \in CS(F)$, $m \in \mathbb{N}$ and $A \in \mathcal{L}_{as}(^mE;F)$, we define $\|A\|_{\alpha,\beta}$ to be $+\infty$ if $A \notin \mathcal{L}_s(^mE_\alpha;F_\beta)$; a similar definition applies to $P \in \mathcal{P}_a(^mE;F)$.

PROPOSITION 25.3 (The Cauchy Inequalities). Let F be separated, $f \in \mathcal{H}(U;F)$, $\alpha \in CS(E)$, $\alpha \neq 0$, $\beta \in CS(F)$, $\rho > 0$, $\xi \in U$ and $\bar{B}_{\alpha,\rho}(\xi) \subset U$. Then

$$\|\frac{1}{m!}\hat{d}^m f(\xi)\|_{\alpha,\beta} \le \frac{1}{\rho^m} \sup_{\alpha(t-\xi)=\rho} \beta[f(t)] \quad \text{for every } m \in \mathbb{N}.$$

PROOF: We apply Proposition 25.2, taking $x \in E$ such that $\alpha(x) = 1$; then

$$\beta[\frac{1}{m!}\hat{d}^m f(\xi)(x)] \le \sup_{\alpha(t-\xi)=\rho} \beta[f(t)].$$

Q.E.D.

PROPOSITION 25.4 Let F be separated, $f \in \mathcal{H}(U;F)$, $\alpha \in CS(E)$, $\beta \in CS(F)$, $\rho > 0$, $\xi \in U$ and $\bar{B}_{\alpha,\beta}(\xi) \subset U$. Then

$$\|\frac{1}{m!}\hat{d}^m f(\xi)\|_{\alpha,\beta} \le \frac{1}{\rho^m} \sup_{\alpha(t-\xi)\le\rho} \beta[f(t)] \quad \text{for every } m \in \mathbb{N}.$$

PROOF: If $\alpha \neq 0$, this reduces to Proposition 25.3. Suppose that $\alpha = 0$. For $m = 0$, the inequality is trivial, so let $m \geq 1$. Since $\alpha = 0$ and $\bar{B}_{\alpha,\rho}(\xi) \subset U$, it follows that U must be E. If $\beta \circ f$ is unbounded, the inequality is trivial. Suppose that $\beta \circ f$ is bounded on E. We shall prove that in this case

$$\beta[\frac{1}{m!} \hat{d}^m f(\xi)(x)] = 0$$

for every $x \in E$. Define $g \in \mathcal{H}(\mathbb{C};F)$ by $g(\lambda) = f(\xi+\lambda x)$. Then $g^{(m)}(0) = d^m f(\xi)(x)$. Hence, applying Proposition 25.3 to g, we have, for every $\beta \in CS(F)$,

$$\beta[\frac{1}{m!} \hat{d}^m f(\xi)(x)] = \beta[\frac{1}{m!} g^{(m)}(0)] \leq \frac{1}{\rho^m} \sup_{t \in E} \beta[f(t)].$$

This holds for every $\rho > 0$, and so, letting $\rho \to \infty$ we obtain the desired result. Q.E.D.

REMARK 25.3: The case $\alpha = 0$ in this proposition is closely related to Liouville's Theorem.

REMARK 25.4: If $\alpha \neq 0$, the constant $1/\rho^m$ in the Cauchy inequality is the best possible constant which is independent of E and F. For, if $P \in \mathcal{P}(^mE;F)$, $\xi \in E$, $f(x) = P(x-\xi)$, then $\frac{1}{m!} \hat{d}^m f(\xi) = P$, and

$$\|P\|_{\alpha,\beta} = \frac{1}{\rho^m} \sup_{\alpha(t-\xi)=\rho} \beta[f(t)] = \frac{1}{\rho^m} \sup_{\alpha(t-\xi)\leq\rho} \beta[f(t)].$$

If $\alpha = 0$, there is no minimal universal constant in Proposition 25.4, since any strictly positive constant could be substituted for $1/\rho^m$.

REMARK 25.5: Under the conditions of Propositions 25.3 and 25.4, we have respectively

$$\left\|\frac{1}{m!}\, d^m f(\xi)\right\|_{\alpha,\beta} \le \frac{m^m}{m!}\,\frac{1}{\rho^m} \sup_{\alpha(t-\xi)=\rho} \beta[f(t)],$$

and

$$\left\|\frac{1}{m!}\, d^m f(\xi)\right\|_{\alpha,\beta} \le \frac{m^m}{m!}\,\frac{1}{\rho^m} \sup_{\alpha(t-\xi)\le\rho} \beta[f(t)].$$

In the first case, $m^m/m!\rho^m$ is the least possible universal constant, while in the case $\alpha = 0$, there is no smallest universal constant.

CHAPTER 26

THE TAYLOR REMAINDER

PROPOSITION 26.1 (The Taylor Remainder Formula). Let F be separated, $f \in \mathcal{H}(U;F)$, $\xi \in U$, $x \in U$ and $\rho > 1$ such that $(1-\lambda)\xi + \lambda x \in U$ for $\lambda \in \mathbb{C}$, $|\lambda| \leq \rho$. Then

$$f(x)-\tau_{m,f,\xi}(x) = \frac{1}{2\pi i}\int_{|\lambda|=\rho} \frac{f[(1-\lambda)\xi+\lambda x]}{\lambda^{m+1}(\lambda-1)}\, d\lambda,$$

where

$$\tau_{m,f,\xi}(x) = \sum_{k=0}^{m} \frac{1}{k!}\, \hat{d}^k f(\xi)(x-\xi), \quad \text{for every } m \in \mathbb{N}.$$

PROOF: By Proposition 25.1,

$$f(x) = \frac{1}{2\pi i}\int_{|\lambda|=\rho} \frac{f[(1-\lambda)\xi+\lambda x]}{\lambda-1}\, d\lambda.$$

By Proposition 25.2, with $x-\xi$ in place of x,

$$\tau_{m,f,\xi}(x) = \frac{1}{2\pi i}\sum_{k=0}^{m}\int_{|\lambda|=\rho} \frac{f[(1-\lambda)\xi+\lambda x]}{\lambda^{k+1}}\, d\lambda.$$

Using the identity

$$\frac{1}{\lambda-1} = \sum_{k=0}^{m}\frac{1}{\lambda^{k+1}} + \frac{1}{\lambda^{m+1}(\lambda-1)},$$

valid for $\lambda \in \mathbb{C}$, $\lambda \neq 0$, $\lambda \neq 1$, we obtain

$$f(x) = \tau_{m,f,\xi}(x) + \frac{1}{2\pi i}\int_{|\lambda|=\rho} \frac{f[(1-\lambda)\xi+\lambda x]}{\lambda^{m+1}(\lambda-1)}\, d\lambda.$$

Q.E.D.

COROLLARY 26.1 Under the conditions of Proposition 26.1, if $\beta \in CS(F)$, then

$$\beta[f(x)-\tau_{m,f,\xi}(x)] \le \frac{1}{\rho^m(\rho-1)} \sup_{|\lambda|=\rho} \beta[f((1-\lambda)\xi+\lambda x)].$$

PROOF: Apply the usual inequalitites for integrals, and note that $|\lambda-1| \ge \rho-1$ for $\lambda \in \mathbb{C}$, $|\lambda| = \rho$. Q.E.D.

COROLLARY 26.2 Under the conditions of Proposition 26.1 we have

$$f(x)-\tau_{m,f,\xi}(x) \in \frac{1}{\rho^m(\rho-1)} X \quad \text{for every} \quad m \in \mathbb{N},$$

where X is the closed convex balanced hull of the set $\{f[(1-\lambda)\xi+\lambda x] : \lambda \in \mathbb{C}, |\lambda| = \rho\}$.

PROOF: Let $V = \{f[(1-\lambda)\xi+\lambda x] : \lambda \in \mathbb{C}, |\lambda| = \rho\}$, and let $\psi \in F'$ be such that $\sup\{|\psi(y)| : y \in V\} \le 1$. Then, by Proposition 26.1,

$$|\psi[f(x)-\tau_{m,f,\xi}(x)]| \le \frac{1}{\rho^m(\rho-1)}.$$

Therefore $f(x) - \tau_{m,f,\xi}(x)$ belongs to the closed convex balanced hull of $\frac{1}{\rho^m(\rho-1)} V$, that is, to $\frac{1}{\rho^m(\rho-1)} X$. Q.E.D.

We have the following converse of Proposition 26.1:

PROPOSITION 26.2 Let F be separated, and $f: U \to F$. In order that $f \in \mathcal{H}(U;F)$ it is necessary and sufficient that f satisfies:

a) f is amply bounded in U.

b) For every $a \in U$ and $b \in E$, the mapping $\lambda \mapsto f(a+\lambda b)$ is continuous in $V = \{\lambda \in \mathbb{C} : a+\lambda b \in U\}$.

c) For every $\xi \in U$ there exist $A_m \in \mathcal{L}_{as}(^mE;F)$ for each $m \in \mathbb{N}$, such that, for every $x \in U$ and $\rho > 1$ for which $(1-\lambda)\xi + \lambda x \in U$ if $\lambda \in \mathbb{C}$, $|\lambda| \le \rho$, we have

$$f(x) - \sum_{k=0}^{m} A_k(x-\xi)^k = \frac{1}{2\pi i}\int_{|\lambda|=\rho} \frac{f[(1-\lambda)\xi+\lambda x]}{\lambda^{m+1}(\lambda-1)}\,d\lambda$$

for every $m \in \mathbb{N}$.

PROOF: Necessity follows from Propositions 23.2 and 26.1.

Sufficiency: We recall that a subset M of E is ξ-balanced if $(1-\lambda)\xi+\lambda x \in M$ for every $x \in M$ and every $\lambda \in \mathbb{C}$, $|\lambda| \leq 1$. By condition a), given $\beta \in CS(F)$, there exists an open ξ-balanced neighbourhood V of ξ in U such that

$$\sup_{x\in V} \beta\{f(x)\} < \infty.$$

It follows from the continuity of the mapping $(\lambda,x) \in \mathbb{C}\times E \mapsto$ $\mapsto (1-\lambda)\xi+\lambda x \in E$, and the fact that V is an open ξ-balanced set that there exists a neighbourhood W of ξ in U, and a number $\rho > 1$, such that $x \in W$, $\lambda \in \mathbb{C}$, $|\lambda| \leq \rho$ implies $(1-\lambda)\xi+\lambda x \in V$. For each $x \in W$ we have, by condition b), that the mapping $\lambda \longmapsto f[\xi+\lambda(x-\xi)] = f[(1-\lambda)\xi+\lambda x]$ is continuous in $\{\lambda \in \mathbb{C} : |\lambda| = \rho\}$, and hence, from c),

$$\beta[f(x) - \sum_{k=0}^{m} A_k(x-\xi)^k] \leq \frac{1}{\rho^m(\rho-1)} \sup_{y\in W} \beta\{f(x)\}.$$

The right hand side of this inequality tends to zero uniformly in W as $m \to \infty$. Therefore, by Proposition 23.2, f is holomorphic. Q.E.D.

PROPOSITION 26.3 Let F be separated, $f \in \mathcal{H}(U)$, $\xi \in U$, and let U_ξ be the largest open ξ-balanced set contained in U. Then for every compact subset K of U_ξ and every

$\beta \in SC(F)$ there exist a neighbourhood V of K in U, and real numbers $c; \theta$, $c \geq 0$, $0 < \theta < 1$, such that

$$\beta[f(x)-\tau_{m,f,\xi}(x)] \leq c\,\theta^m \quad \text{for every} \quad x \in V, \quad m \in \mathbb{N}.$$

PROOF: The mapping

$$(\lambda,x) \longmapsto \beta[f((1-\lambda)\xi+\lambda x)] \in \mathbb{R}$$

is defined and continuous in the open subset W of $\mathbb{C} \times E$ consisting of all (λ,x) for which $(1-\lambda)\xi+\lambda x \in U$. If D is the closed unit disc in $\mathbb{C}$, then since $K \subset U_\xi$, the compact set $D \times K$ lies in W. Therefore the above mapping is bounded on $D \times K$. It follows that there is a neighbourhood V of K in U and $\rho > 1$ such that

$$\sup\{\beta[f((1-\lambda)\xi+\lambda x] : x \in V,\ \lambda \in \mathbb{C},\ |\lambda| \leq \rho\} = B < \infty.$$

The desired inequality now follows from Corollary 26.1, with $\theta = 1/\rho$ and $C = B/(\rho-1)$. Q.E.D.

DEFINITION 26.1 Let $\mathfrak{X}$ be a set of mappings of U into F. $\mathfrak{X}$ is said to be amply bounded in U if for every $\beta \in CS(F)$ and $\xi \in U$ there exists a neighbourhood V of ξ in U such that

$$\sup_{f\in\mathfrak{X},\, x\in V} \beta\{f(x)\} < \infty.$$

REMARK 26.1: In practice, Proposition 26.3 is often applied in the following form: under the conditions of the Proposition, if $\{\lambda_m\}_{m\in\mathbb{N}}$ is a sequence of complex numbers such that $\limsup |\lambda_m|^{1/m} \leq 1$, then the sequence

$$\lambda_m(f-\tau_{m,f,\xi})/U_\xi \,, \qquad m \in \mathbb{N},$$

is amply bounded in U_ξ.

REMARK 26.2: This result can be generalized as follows: If F is separated, $\mathfrak{X} \subset \mathcal{H}(U;F)$ is amply bounded, $\xi \in U$ and $\limsup |\lambda_m|^{1/m} \le 1$, then the family of mappings

$$\lambda_m(f-\tau_{m,f,\xi})/U_\xi \,, \qquad f \in \mathfrak{X}, \quad m \in \mathbb{N}$$

is amply bounded in U_ξ.

REMARK 26.3: Let F be separated, $f \in \mathcal{H}(U;F)$, $\xi \in U$, $\alpha \in CS(E)$, $\alpha \neq 0$, $r > 0$, $\bar{B}_{\alpha,r}(\xi) \subset U$, $x \in B_{\alpha,r}(\xi)$, $\alpha(x-\xi) \neq 0$, and $\beta \in CS(F)$. Then

$$\frac{\beta[f(x)-\tau_{m,f,\xi}(x)]}{[\alpha(x-\xi)]^{m+1}} \le \frac{1}{r^m[r-\alpha(x-\xi)]} \sup_{\alpha(t-\xi)=r} \beta[f(t)],$$

for every $m \in \mathbb{N}$.

To see this, we apply Corollary 26.1 with $\rho = r/\alpha(x-\xi)$, noting that $\alpha[(1-\lambda)\xi+\lambda x-\xi] = |\lambda|\,\alpha(x-\xi) = r$ when $|\lambda| = \rho$.

CHAPTER 27

COMPACT AND LOCAL CONVERGENCE OF THE TAYLOR SERIES

We recall that if $\xi \in U$, then U_ξ denotes the largest ξ-balanced subset of U. Thus

$$U_\xi = \{x \in U : (1-\lambda)\xi + \lambda x \in U \text{ for every } \lambda \in \mathbb{C}, |\lambda| \leq 1\}.$$

It is easy to see that U_ξ is open.

PROPOSITION 27.1 Let F be separated, $f \in \mathcal{H}(U;F)$ and $\xi \in U$. Then for every compact subset K of U_ξ and every $\beta \in CS(F)$ there exists a neighbourhood V of K in U such that

$$\lim_{m\to\infty} \beta[f(x) - \tau_{m,f,\xi}(x)] = 0$$

uniformly in $x \in V$.

PROOF: By Proposition 26.3 there is a neighbourhood V of K in U, and real numbers c, θ, where $c \geq 0$ and $0 < \theta < 1$, such that

$$\beta[f(x) - \tau_{m,f,\xi}(x)] \leq c\cdot\theta^m \quad \text{for every } x \in V, \ m \in \mathbb{N}.$$

The proposition follows immediately. Q.E.D.

PROPOSITION 27.2 Let F be separated, $f \in \mathcal{H}(U;F)$ and $\xi \in U$. Then

$$f(x) = \sum_{m=0}^{\infty} \frac{1}{m!} \hat{d}^m f(\xi)(x-\xi)$$

uniformly on every compact subset of U_ξ.

PROOF: This follows immediately from Proposition 27.1.

Q.E.D.

COROLLARY 27.1 (Unique Determination of a Holomorphic Mapping by its Taylor Series). Let F be separated, $\xi \in U$ and $f,g \in \mathcal{H}(U;F)$. If $d^m f(\xi) = d^m g(\xi)$ for every $m \in \mathbb{N}$ then $f = g$ in U_ξ.

PROPOSITION 27.3 Let F be separated, $\xi \in U$, $f \in \mathcal{H}(U;F)$, and suppose also that f is locally bounded. Then

$$f(x) = \sum_{m=0}^{\infty} \frac{1}{m!} \hat{d}^m f(\xi)(x-\xi)$$

uniformly on some neighbourhood in U of every compact subset of U_ξ.

PROOF: It suffices to show that the Taylor series of f at ξ converges uniformly on a neighbourhood of each $x \in U_\xi$. We claim that if $x \in U_\xi$ there exists a neighbourhood W of x in U and $\rho > 1$ such that the set

$$T = \{(1-\lambda)\xi+\lambda t : \lambda \in \mathbb{C}, |\lambda| \leq \rho, t \in W\}$$

is contained in U, and $f(T)$ is bounded in F.

To see this, consider the continuous mapping $G: (\lambda,t) \in \mathbb{C}\times E \longmapsto (1-\lambda)\xi+\lambda t \in E$.

If $x \in U_\xi$, the compact set $M = \{(1-\lambda)\xi+\lambda x : \lambda \in \mathbb{C}, |\lambda| \leq 1\}$ is contained in U_ξ. Since f is locally bounded, there is a neighbourhood U' of M in U such that $f(U')$ is bounded in F. Since $G^{-1}(U')$ is a neighbourhood of

$\bar{B}_1(0) \times \{x\}$ in $\mathbb{C} \times E$ there exist a neighbourhood W of x in U and $\rho > 1$ such that $(1-\lambda)\xi + \lambda x \in U'$ if $\lambda \in \mathbb{C}$, $|\lambda| \leq \rho$ and $t \in W$. This establishes our claim.

It follows from Corollary 26.1 that if $\beta \in CS(F)$, $m \in \mathbb{N}$ and $t \in W$, then

$$\beta[f(t) - \tau_{m,f,\xi}(t)] \leq \frac{1}{\rho^m(\rho-1)} \sup_{t' \in T} \beta[f(t')].$$

Since V is independent of β, it follows that the Taylor series of f at ξ converges to $f(t)$ uniformly in $t \in W$.

Q.E.D.

REMARK 27.1: If $f \in \mathcal{P}(E;F)$ and F is separated, the Taylor series of f at any point of E contains only a finite number of non-zero terms, and hence converges to f uniformly in E, whether or not f is locally bounded. A simple example shows that f need not be locally bounded: If E is not seminormable then the identity mapping $I: E \to E$ is not locally bounded, but $I \in \mathcal{P}(E;E)$.

REMARK 27.2: There are two interesting situations in which every $f \in \mathcal{H}(U;F)$ is locally bounded; one is where E is finite dimensional, the other where F is seminormed.

We have the following partial converse to Proposition 27.3:

PROPOSITION 27.4 Let E be seminormable, F separated, and $f \in \mathcal{H}(U;F)$. If the Taylor series of f at a point ξ of U converges to f uniformly in some neighbourhood of ξ, then f is locally bounded at ξ.

PROOF: Since E is seminormable, there is a bounded neighbourhood V of ξ in U such that

$$f(x) = \sum_{m=0}^{\infty} \frac{1}{m!} d^m f(\xi)(x-\xi)^m$$

uniformly in $x \in V$. Continuous polynomials are bounded on bounded sets, and hence $\beta \circ f$ is bounded on V for every $\beta \in CS(F)$. Therefore, since V is independent of β, f is bounded on V. Q.E.D.

REMARK 27.3: The requirement that E be seminormable in Proposition 27.4 is essential, for if E is not seminormable, the identity $I: E \to E$ is not locally bounded.

We now present two examples of a holomorphic mapping whose Taylor series at a point ξ does not converge uniformly in any neighbourhood of ξ. In Example 27.1 E and F are Fréchet-Montel spaces if I is countable, and in Example 27.2 E is a normed space and F a Fréchet-Montel space.

EXAMPLE 27.1 Let I be a non-empty set, and let $E = F = \mathbb{C}^I$. Let $g \in \mathcal{H}(\mathbb{C})$, and define $f \in \mathcal{H}(E;F)$ by

$$x = (x_j)_{j\in I} \in E \longmapsto f(x) = (g(x_j))_{j\in I} \in F.$$

If I is infinite, and g is not a polynomial, the Taylor series of f at any point $\xi \in E$ does not converge uniformly in any neighbourhood of ξ. Suppose this were not so. Then, for some $\xi = (\xi_j)_{j\in I} \in E$ there exists a finite subset J of I and $\varepsilon > 0$ such that the Taylor series of f at ξ converges uniformly in $V = \{x = (x_j)_{j\in I} \in E : |x_j-\xi_j| < \varepsilon$

for $j \in J\}$. Let $k \in I\setminus J$ and let $\beta \in SC(F)$ be defined by $\beta(y) = |y_k|$, $y = (y_j)_{j\in I} \in F$. Then, applying β to the Taylor series of f at ξ, we find that the series

$$\sum_{m=0}^{\infty} a_m(x_k-\xi_k)^m$$

is uniformly convergent for $x_k \in \mathbb{C}$, where $a_m = \frac{1}{m!} g^{(m)}(\xi_k)$. Hence there exists $n \in \mathbb{N}$ such that

$$|a_m(x_k-\xi_k)^m| \leq 1 \text{ for every } m \geq n, \quad x_k \in \mathbb{C}.$$

But this implies that $a_m = 0$ for $m \geq n$, which means that g is a polynomial, contrary to our hypothesis.

EXAMPLE 27.2 Let E be a complex normed space of infinite dimension. Then, by Proposition 24.4, there exists $g \in \mathcal{H}(E)$ which is unbounded on some bounded subset of E. We define $g_n \in \mathcal{H}(E)$, $n \in \mathbb{N}$, by $g_n(x) = g(nx)$, $x \in E$. Now let $F = \mathbb{C}^{\mathbb{N}}$, and let $f \in \mathcal{H}(E;F)$ be given by

$$x \in E \longmapsto f(x) = (g_n(x))_{n\in\mathbb{N}} \in F.$$

Suppose that for some $r > 0$ the Taylor series of f at the origin converges uniformly in the open ball, V, of radius r. Then

$$\sum_{k=0}^{\infty} \frac{1}{k!} d^k g_n(0)x^k$$

converges uniformly to g_n in V for every $n \in N$. This implies that

$$\sum_{k=0}^{\infty} \frac{1}{k!} d^k g(0)x^k$$

converges uniformly to g in nV for every $n \in \mathbb{N}$. Since every continuous polynomial is bounded on the bounded set nV for every $n \in \mathbb{N}$, it follows that g is bounded on nV for every $n \in \mathbb{N}$. Therefore g is bounded on every bounded subset of E, which contradicts our choice of g. Therefore f is not locally bounded at the origin.

REMARK 27.4: Given E, $\xi \in U$, a neighbourhood V of ξ in U, and a separated space F, one can ask whether there exists $f \in \mathcal{H}(U;F)$ whose Taylor series at ξ does not converge uniformly in V. A solution to this problem when $U = E$, $\xi = 0$ and $F = \mathbb{C}$ implies a general solution.

PROPOSITION 27.5 If E is not seminormable (hence the dimension of E is infinite) and $F \neq \{0\}$ is separated, then for every neighbourhood V of the origin in E there exists $f \in \mathcal{H}(E;F)$ whose Taylor series at the origin does not converge uniformly in V.

PROOF: Choose $\alpha_1 \in CS(E)$ such that $B_{\alpha_1,1}(0) \subset V$. Since E is not seminormable, there exists $\alpha_2 \in CS(E)$ such that $\alpha_2 \leq c\alpha_1$ is false for all $c \in \mathbb{R}_+$. Hence there exists $\varphi \in (E_{\alpha_2})'$ such that $\varphi \notin (E_{\alpha_1})'$. Choose $g \in \mathcal{H}(\mathbb{C})$ so that g is not a polynomial, and $b \in F$, $b \neq 0$. Then $f = (g \circ \varphi) \cdot b \in \mathcal{H}(E_{\alpha_2};F) \subset \mathcal{H}(E;F)$, and the Taylor series of f at the origin is

$$\sum_{m=0}^{\infty} a_m \varphi^m \cdot b, \quad \text{where} \quad a_m = \frac{1}{m!} g^{(m)}(0).$$

If this series converges uniformly in V, then, choosing $\beta \in CS(F)$ such that $\beta(b) = 1$, there exists $n \in \mathbb{N}$ such that

$$\beta\{a_m[\varphi(x)]^m \cdot b\} \leq 1 \quad \text{for} \quad x \in V, \quad m \geq n.$$

Therefore

$$|\varphi(x)| \leq \frac{1}{|a_m|^{1/m}} \quad \text{for} \quad x \in V, \quad \text{and} \quad m \geq n \quad \text{such that } a_m \neq 0.$$

Since g is not a polynomial, $a_m \neq 0$ for infinitely many $m \in \mathbb{N}$. In particular, there exists $m \geq n$ for which $a_m \neq 0$. Since $B_{\alpha_1,1}(0) \subset V$, this implies that φ is continuous with respect to α_1, which is a contradiction. Q.E.D.

PROPOSITION 27.6 Let E be separated, and let U be a non-empty open subset of E. The following are equivalent:

a) For every separated space F, and every $f \in \mathcal{H}(U;F)$, the Taylor series of f at each point ξ of U converges uniformly in some neighbourhood of ξ.

b) The dimension of E is finite.

PROOF: a) $\Rightarrow$ b). Suppose first that E is not normable. Let I be a base of neighbourhood of the origin in E. For each $V \in I$ there exists, by Proposition 27.5, $f_V \in \mathcal{H}(E)$ whose Taylor series at the origin does not converge uniformly in V. Let $F = \mathbb{C}^I$, and define $f \in \mathcal{H}(E;F)$ by

$$x \in E \longmapsto f(x) = (f_V(x))_{V \in I} \in F.$$

Then, as in Example 27.1, the Taylor series of f at the origin does not converge uniformly in any neighbourhood of the origin.

Now suppose that E is normable and has infinite dimension, and let $F = \mathbb{C}^{\mathbb{N}}$. Example 27.2 shows that there exists $f \in \mathcal{H}(E;F)$ whose Taylor series at the origin does not converge in any neighbourhood of the origin.

b) $\Rightarrow$ a) follows from Proposition 27.3, since every finite dimensional space is locally compact. Q.E.D.

REMARK 27.3: The proof of this proposition relies on the Josefson-Nissenzweig Theorem (Proposition 24.3) when E is normed. The following result shows that this theorem is an essential part of the proof:

If E is a normed space, the following are equivalent:

a) For every F, and every $f \in \mathcal{H}(U;F)$, the Taylor series of f at ξ converges uniformly in some neighbourhood of ξ.

b) Every $f \in \mathcal{H}(E)$ is bounded on every bounded subset of E.

To see this suppose a) holds. Then, by Proposition 27.6, E is finite dimensional and hence every bounded subset of E is relatively compact. Therefore a) implies b).

Conversely, suppose that every $f \in \mathcal{H}(E)$ is bounded on every bounded subset of E. It follows that for every F, every $f \in \mathcal{H}(E;F)$ is bounded on every bounded subset of E. In particular, f is bounded on every bounded neighbourhood of the origin, and it follows from the Cauchy integral formula that the Taylor series of every $f \in \mathcal{H}(E;F)$ converges uniformly on some neighbourhood of the origin, from which a) follows.

Without the Josefson-Nissenzweig Theorem, one would be unable to show that for every infinite dimensional complex normed space E there exists $f \in \mathcal{H}(E)$ which is not bounded on every bounded ball.

CHAPTER 28

THE MULTIPLE CAUCHY INTEGRAL AND THE CAUCHY INEQUALITIES

PROPOSITION 28.1 (The Multiple Cauchy Integral). Let F be separated, $f \in \mathcal{H}(U;F)$, $\xi \in U$, $k \in \mathbb{N}^*$, $x_1,\ldots,x_k \in E$, $n_1,\ldots,n_k \in \mathbb{N}$, $m = n_1 +\ldots+ n_k$ and $\rho_1,\ldots,\rho_k > 0$, such that $\xi + \lambda_1 x_1 +\ldots+ \lambda_k x_k \in U$ for every $\lambda_j \in \mathbb{C}$, $|\lambda_j| \le \rho_j$, $j = 1,\ldots,k$. Then

$$\frac{1}{n_1!\ldots n_k!} \hat{d}^m f(\xi) x_1^{n_1} \ldots x_k^{n_k} =$$

$$= \frac{1}{(2\pi i)^k} \int_{\substack{|\lambda_j|=\rho_j \\ 1\le j\le k}} \frac{f(\xi+\lambda_1 x_1+\ldots+\lambda_k x_k)}{\lambda_1^{n_1+1}\ldots\lambda_k^{n_k+1}} \, d\lambda_1 \ldots d\lambda_k \, .$$

REMARK 28.1: The proof of Proposition 28.1 is similar to the proof of Proposition 25.2, the single integral in the latter case being replaced by a multiple integral. Alternatively, Proposition 28.1 can be obtained by repeated application of Proposition 25.2. Proposition 25.2 is the extreme case of Proposition 28.1 in which $k = 1$. The other extreme case, where $k = m$, $n_1 =\ldots= n_m = 1$ is as follows:

COROLLARY 28.1 Let F be separated, $f \in \mathcal{H}(U;F)$, $\xi \in U$, $m \in \mathbb{N}^*$, $x_1,\ldots,x_m \in E$ and $\rho_1,\ldots,\rho_m > 0$ such that $\xi + \lambda_1 x_1 +\ldots+ \lambda_m x_m \in U$ for every $\lambda_j \in \mathbb{C}$, $|\lambda_j| \le \rho_j$, $j = 1,\ldots,m$. Then

$$d^m f(\xi)(x_1,\dots,x_m) =$$

$$= \frac{1}{(2\pi i)^m} \int_{\substack{|\lambda_j|=\rho_j \\ 1\le j\le m}} \frac{f(\xi+\lambda_1 x_1+\dots+\lambda_m x_m)}{(\lambda_1\dots\lambda_m)^2}\, d\lambda_1\dots d\lambda_m$$

REMARK 28.2: If we write $n = (n_1,\dots,n_k) \in \mathbb{N}^k$, $d(n) = k$, $|n| = n_1 +\dots+ n_k$, $n! = n_1!\dots n_2!$, $x = (x_1,\dots,x_k) \in E^k$, $\lambda = (\lambda_1,\dots,\lambda_k) \in \mathbb{C}^k$, $\rho = (\rho_1,\dots,\rho_k)$, $x^n = x_1^{n_1}\dots x_k^{n_k}$, $\lambda x = \lambda_1 x_1 +\dots+ \lambda_k x_k$, $\lambda^{n+1} = \lambda_1^{n_1+1}\dots\lambda_k^{n_k+1}$, $d\lambda = d\lambda_1\dots d\lambda_n$, and if $|\lambda_1| = \rho_1,\dots,|\lambda_k| = \rho_k$ is written $|\lambda| = \rho$, then integral formula given in Proposition 28.1 becomes

$$\frac{1}{n!} d^{|n|} f(\xi)x^n = \frac{1}{(2\pi i)^{d(n)}} \int_{|\lambda|=\rho} \frac{f(\xi+\lambda x)}{\lambda^{n+1}}\, d\lambda,$$

a form similar to Proposition 25.2.

REMARK 28.3: Cauchy inequalities can be derived from Proposition 28.1, or from Remark 28.2, in the same way as the Cauchy inequalities of Chapter 25 were derived from Proposition 22.2.

REMARK 28.4: If we apply Corollary 28.1 to $f = \hat{A}$, where $A \in \mathcal{L}_s(^mE;F)$, with $\xi = 0$ and $U = E$, we obtain a new polarization formula:

$$A(x_1,\dots,x_m) = \frac{1}{m!(2\pi i)^m} \int_{\substack{|\lambda_j|=1 \\ 1\le j\le m}} \frac{\hat{A}(\lambda_1 x_1+\dots+\lambda_m x_m)}{(\lambda_1\dots\lambda_m)^2}\, d\lambda_1\dots d\lambda_m,$$

where we have taken $\rho_1 =\dots= \rho_m = 1$; in this case, it can be shown easily by a change of variable that any choice of

$\rho_1,\ldots,\rho_m > 0$ gives the same value for the integral. Like the original polarization formula (Lemma 15.1), we can use this formula to obtain an estimate for $\beta[A(x_1,\ldots,x_m)]$; if $\alpha(x_j) \leq 1$ for $j = 1,\ldots,m$, then

$$\beta[A(x_1,\ldots,x_m)] \leq \frac{m^m}{m!} \|\hat{A}\|_{\alpha,\beta} .$$

REMARK 28.5: More generally, if we apply Proposition 28.1 to $f = \hat{A}$, where $A \in \mathcal{L}_s(^mE;F)$, $m \in \mathbb{N}^*$, $\xi = 0$ and $U = E$, we obtain another new polarization formula:

$$\frac{1}{n_1!\ldots n_k!} Ax_1^{n_1}\ldots x_k^{n_k} =$$

$$= \frac{1}{m!(2\pi i)^k} \int_{\substack{|\lambda_j|=1 \\ 1\leq j\leq k}} \frac{\hat{A}(\lambda_1 x_1+\ldots+\lambda_k x_k)}{\lambda_1^{n_1+1}\ldots\lambda_k^{n_k+1}} d\lambda_1\ldots d\lambda_k .$$

CHAPTER 29

DIFFERENTIALLY STABLE SPACES

DEFINITION 29.1 Let F be a complex locally convex space. F is said to be differentially stable if for every E and every non-empty open subset U of E, $\mathcal{H}(U;F) = H(U;F)$. We recall that $H(U;F) = \mathcal{H}(U;\hat{F}) \cap F^U$. Thus if F is separated, F is differentially stable if and only if, for every $f \in H(U;F)$, we have $d^m f(\xi)(x_1,\ldots,x_m) \in F$ for every $m \in N^*$, $\xi \in U$, and $x_1,\ldots,x_m \in E$. Bearing in mind the proof of the Cauchy integral formula (Proposition 25.2), it suffices to consider the case $E = \mathbb{C}$, $U = B_1(0)$, and to show that $f^{(m)}(0) \in F$ for every $f \in \mathcal{H}(U;F)$.

It is easy to see that a complete space is differentially stable. The following proposition describes two more general conditions, either of which guarantees differential stability.

PROPOSITION 29.1 F is differentially stable in each of the following cases:

1) F is sequentially complete.

2) The closed convex balanced hull of every compact subset of F is compact.

PROOF: It suffices to consider the case where F is separated.

Let $f \in H(U;F)$, $\xi \in U$, and $x \in E$. Choose $\rho > 0$ such that $(1-\lambda)\xi+\lambda x \in U$ for $\lambda \in \mathbb{C}$, $|\lambda| \le \rho$. Then, by Proposition 25.2,

$$\frac{1}{m!} d^m f(\xi)x^m = \frac{1}{2\pi i}\int_{|\lambda|=\rho} \frac{f(\xi+\lambda x)}{\lambda^{m+1}} d\lambda \quad \text{for every} \quad m \in \mathbb{N}.$$

Since $f(\xi+\lambda x) \in F$ for every ξ, λ, x, conditions 1) and 2) each imply that the integral which appears in this equation takes its values in F. Hence, by the polarization formula, $f \in \mathcal{H}(U;F)$. Q.E.D.

PROPOSITION 29.2 F is differentially stable if and only if the space $WF = (F, \sigma(F,F'))$ is differentially stable.

In order to prove this proposition we need the following result from the theory of weakly holomorphic mappings.

LEMMA 29.1 Let $E = \mathbb{C}$, and $f: U \to F$. Then $f \in H(U;F)$ if and only if $\psi \circ f \in H(U)$ for every $\psi \in F'$.

PROOF: The necessity of this condition is obvious. To prove its sufficiency, suppose that $\psi \circ f \in H(U)$ for every $\psi \in F'$, suppose that F is separated, and let $\xi \in U$. We show first that f is continuous at ξ. If $g \in \mathcal{H}(U)$ and $\bar{B}_\rho(\xi) \subset U$, then for $z \in B_\rho(\xi)$ we have

$$g(z) = \frac{1}{2\pi i}\int_{|t-\xi|=\rho} \frac{g(t)}{t-z} dt,$$

$$g(\xi) = \frac{1}{2\pi i}\int_{|t-\xi|=\rho} \frac{g(t)}{t-\xi} dt,$$

and hence

$$g(z)-g(\xi) = \frac{z-\xi}{2\pi i} \int_{|t-\xi|=\rho} \frac{g(t)}{(t-z)(t-\xi)}\, dt.$$

Therefore

$$|g(z)-g(\xi)| \le \frac{|z-\xi|}{\rho-|z-\xi|} \sup_{|t-\xi|=\rho} |g(t)|.$$

Applying this to $g = \psi \circ f$, where $\psi \in F'$, we have

$$|\psi[f(z)-f(\xi)]| \le \frac{|z-\xi|}{\rho-|z-\xi|} \sup_{|t-\xi|=\rho} |\psi[f(t)]|.$$

Since this holds for every $\psi \in F'$, it follows that

$$\beta[f(z)-f(\xi)] \le \frac{|z-\xi|}{\rho-|z-\xi|} \sup_{|t-\xi|=\rho} \beta[f(t)]$$

for every $\beta \in CS(F)$. It is easy to see that f is bounded on every compact set, and hence $\beta[f(z)-f(\xi)] \to 0$ as $z \to \xi$. Therefore f is continuous.

We have shown that

$$(\psi \circ f)(z) = \frac{1}{2\pi i} \int_{|t-\xi|=\rho} \frac{(\psi \circ f)(t)}{t-z}\, dt$$

for every $\psi \in F'$. Since F is separated, it follows that

$$f(z) = \frac{1}{2\pi i} \int_{|t-\xi|=\rho} \frac{f(t)}{t-z}\, dt,$$

where $z \in B_\rho(\xi)$. We have

$$\frac{1}{t-z} = \frac{1}{(t-\xi)-(z-\xi)} = \sum_{m=0}^{\infty} \frac{(z-\xi)^m}{(t-\xi)^{m+1}}$$

for $|t-\xi| = \rho$ and $z \in B_\rho(\xi)$. Furthermore, the convergence is uniform for $|t-\xi| = \rho$ and $z \in \bar{B}_r(\xi)$, where $0 < r < \rho$.

It follows that

$$f(z) = \sum_{m=0}^{\infty} \left[\frac{1}{2\pi i} \int_{|t-\xi|=\rho} \frac{f(t)}{(t-\xi)^{m+1}} dt\right](z-\xi)^m$$

uniformly on every compact subset of $B_\rho(\xi)$.

Therefore $f \in \mathcal{H}(U;\hat{F})$, and since f maps U into F, it follows that $f \in H(U;F)$. Q.E.D.

PROOF OF PROPOSITION 29.2: Suppose that F is differentially stable. Let U be a non-empty open subset of $\mathbb{C}$ and let $f \in H(U,WF)$. Then $\psi \circ f \in H(U)$ for every $\psi \in F'$, and so, by Lemma 29.1, $f \in H(U;V) = \mathcal{H}(U;F)$. Since the topology $\sigma(F,F')$ is weaker than the topology of F, $\mathcal{H}(U;F) \subset \mathcal{H}(U,WF)$, and hence $f \in \mathcal{H}(u,WF)$. Therefore WF is differentially stable.

Conversely, suppose that WF is differentially stable, and let $f \in H(U;F)$, where U is a non-empty open subset of $\mathbb{C}$. Then $f \in H(U;WF) = \mathcal{H}(U;WF)$. Let $\xi \in U$, and let $\hat{F}$ be a completion of F. If $\psi \in (\hat{F})'$, then

1) Since $\psi \circ f$ is the composition of $f \in H(U;F)$ with ψ, $(\psi \circ f)^{(m)}(\xi) = \psi[f^{(m)}(\xi)]$ for every $m \in \mathbb{N}$.

2) Since $\psi \circ f$ is the composition of $f \in \mathcal{H}(U;WF)$ with ψ, $(\psi \circ f)^{(m)}(\xi) = \psi[f_W^{(m)}(\xi)]$ for every $m \in \mathbb{N}$, where $f_W^{(m)}(\xi) \in F$ is the differential or order m of f at ξ, taking f as a holomorphic mapping of U into WF.

Therefore $f^{(m)}(\xi) = f_W^{(m)}(\xi) \in F$ for every $m \in \mathbb{N}$, $\xi \in U$, which implies that F is differentially stable.

Q.E.D.

CHAPTER 30

LIMITS OF HOLOMORPHIC MAPPINGS

PROPOSITION 30.1 $H(U;F)$ is closed in $C(U;F)$ in the compact-open topology.

PROOF: We recall that the compact-open topology on $C(U;F)$ is defined by the family of seminorms

$$g \in C(U;F) \longmapsto \sup_{x \in K} \beta\{g(x)\}$$

where $\beta \in CS(F)$ and $K \subset U$ is compact. It suffices to consider the case where F is separated. Let f belong to the closure of $H(U;F)$ in $C(U;F)$ in the compact-open topology. Fixing $\xi \in U$, let $A_o = f(\xi)$. We define a mapping $A_m: E^m \to F$ for each $m \in \mathbb{N}^*$ as follows: if $x_1,\ldots,x_m \in E$, choose real numbers $\rho_1,\ldots,\rho_m > 0$ such that $\xi + \lambda_1 x_1 + \ldots + \lambda_m x_m \in U$ for $\lambda_j \in \mathbb{C}$, $|\lambda_j| \leq \rho_j$, $j = 1,\ldots,m$, and let

$$A_m(x_1,\ldots,x_m) = \frac{1}{m!(2\pi i)^m} \int_{\substack{|\lambda_j|=\rho_j \\ 1\leq j\leq m}} \frac{f(\xi+\lambda_1 x_1+\ldots+\lambda_m x_m)}{(\lambda_1\ldots\lambda_m)^2} d\lambda_1\ldots d\lambda_m$$

We claim that:

1) The value of $A_m(x_1,\ldots,x_m)$ is independent of the choice of $\rho_1,\ldots,\rho_m$ satisfying the stated condition.

2) Each A_m is m-linear and symmetric.

3) If $x \in U$, and $\rho > 1$ is a real number such that $(1-\lambda)\xi+\lambda x \in U$ for $\lambda \in \mathbb{C}$, $|\lambda| \leq \rho$, then

$$f(x) - \sum_{k=0}^{m} A_k(x-\xi)^k = \frac{1}{2\pi i} \int_{|\lambda|=\rho} \frac{f[(1-\lambda)\xi+\lambda x]}{\lambda^{m+1}(\lambda-1)} d\lambda .$$

If $f \in H(U;F)$, then $A_m = \frac{1}{m!} d^m f(\xi)$, and these properties follow from the Cauchy integral formula (Proposition 25.2) and the Taylor remainder formula (Proposition 26.1). Using the same methods as were used to prove Proposition 4.6, Part I, it follows that 1), 2) and 3) hold when f belongs to the closure of $H(U;F)$ in $C(U;F)$. It now follows from Proposition 26.2 that $f \in H(U;F)$. Q.E.D.

REMARK 30.1: In the proof of this proposition, the fact that f is continuous implies that each A_m is continuous. It then follows from 3), using the methods of proof of Proposition 26.2, that $f \in H(U;F)$, without appealing to Proposition 26.3.

REMARK 30.2: The proof of Proposition 30.1 also shows that $H(U;F)$ is closed in $C(U;F)$ in the topology of uniform convergence on the finite dimensional compact subsets of U. (If E is separated and has infinite dimension, and $F \neq \{0\}$, this topology is strictly weaker than the compact-open topology on $H(U;F)$).

REMARK 30.3: It is possible to avoid the use of multiple integrals in the proof of Proposition 30.1. Hence $H(U;F)$ is closed in $C(U;F)$ in the topology of uniform convergence on one dimensional compact subsets of U.

REMARK 30.4: If $C(U;F)$ is complete in the compact-open topology (the case most frequently encountered is that in which E is semimetrizable and F is complete) then $H(U;F)$ is complete in the compact-open topology. However, $C(U;F)$ need not be complete in the compact-open topology, even when $F = \mathbb{C}$. Furthermore, $H(U;F)$ need not be complete in this topology even when $F = \mathbb{C}$.

EXAMPLE 30.1 Let E be $\mathbb{C}^{(I)}$, where I has at least the power of the continuum. We place on E the finest locally convex topology. This is the locally convex inductive limit topology corresponding to the decomposition $E = \bigcup S$, where S ranges over the set of finite dimensional vector subspaces of E. There exists a 2-homogeneous polynomial, P, from E into $\mathbb{C}$ which is not continuous for this topology (see [11], pages 42-44). There exist unique $c_{ij} \in \mathbb{C}$, $c_{ij} = c_{ji}$, $i,j \in I$, such that

$$P(x) = \Sigma\, c_{ij}\, x_i\, x_j\,,$$

where $x = (x_i)_{i\in I} \in E$. For each finite $J \subset I$, let $P_J\colon E \to \mathbb{C}$ be defined by

$$P_J(x) = \sum_{i,j\in J} c_{ij}\, x_i\, x_j\,.$$

Then $P_J \in \mathcal{P}(^2E) \subset H(U)$, and $P_J \to P$ uniformly on the compact subsets of E. However, $P \notin H(U)$; thus $H(U)$ is not complete in the compact-open topology, where U is any non-empty open subset of E.

CHAPTER 31

UNIQUENESS OF HOLOMORPHIC CONTINUATION

PROPOSITION 31.1 Let $f \in \mathcal{H}(U;F)$, where F is separated and U is connected. Then

a) f is identically zero in U if and only if f is identically zero in some non-empty open subset of U.

b) f is identically zero in U if and only if there exists $\xi \in U$ such that $d^m f(\xi) = 0$ for every $m \in \mathbb{N}$.

PROOF: We note first that if f is identically zero in a non-empty open subset V of U then $\hat{d}^m f$ is also identically zero in V for every $m \in \mathbb{N}$. This follows immediately from the uniqueness of the Taylor series of f at each point. Moreover, if $\xi \in U$ is such that $d^m f(\xi) = 0$ for every $m \in \mathbb{N}$, then f is identically zero in the largest ξ-balanced subset U_ξ of U, which is necessarily open. This follows from the fact that the Taylor series of f at ξ converges at every point of U_ξ.

The necessity of the condition given in a) is obvious. To prove its sufficiency, suppose that f is identically zero in some non-empty open subset of U. Let V be the set of points in U at which the value of f is not zero. Then V is a non-empty open subset of U. It follows from our first remark above that $\hat{d}^m f$ is identically zero in V for every $m \in N$. Now, by the Cauchy integral formula, it is easy to see that for each $m \in \mathbb{N}^*$ and $x \in E$, the mapping $t \in U \longmapsto \hat{d}^m f(t)(x)$ is continuous. Therefore $\hat{d}^m f$ is identically zero in the closure, $\bar{V}$, of V in U. Hence $d^m f$ is also identically zero in $\bar{V}$. By our second remark, f is then identically zero in a neighbourhood of $\bar{V}$ in U. Therefore $\bar{V} \subset V$, which implies that V is closed. Since U is connected, we conclude that $V = U$, that is, f vanishes throughout U.

The necessity of the condition given in b) follows from the first remark. Conversely, suppose there exists $\xi \in U$ such that $d^m f(\xi) = 0$ for every $m \in \mathbb{N}$. By the second remark, f is then identically zero in some neighbourhood of ξ, and hence, by a), f is identically zero in U. Q.E.D.

PROPOSITION 31.2 Let $f \in \mathcal{H}(U;F)$, where U is connected. Then

a) For every non-empty open subset V of U the closed vector subspaces of F generated by $f(V)$ and $f(U)$ respectively are equal.

b) If F is separated, then for every $\xi \in U$ the closed vector subspaces of F generated by $\{d^m f(\xi)x^m : x \in E, m \in \mathbb{N}\}$ and $f(U)$ respectively are equal.

PROOF: a) Let S_U and S_V be the closed vector subspaces of F generated by $f(U)$ and $f(V)$ respectively. Let π_V be the canonical mapping of F onto the separated locally convex quotient space F/S_V. Then $\pi_V \circ f \in \mathcal{H}(U;F/S_V)$, and

$\pi_V \circ f$ is identically zero in V. Hence, by Proposition 31.1, $\pi_V \circ f$ is identically zero in U. Therefore $f(U) \subset S_V$, which implies that $S_U \subset S_V$. Moreover, $S_V \subset S_U$, since $f(V) \subset \subset f(U)$. Therefore $S_U = S_V$.

b) Let T_ξ be the closed vector subspace of F generated by $\{d^m f(\xi) x^m : m \in \mathbb{N}, x \in E\}$. The holomorphic mapping $\pi_U \circ f: U \to F/S_U$ is identically zero in U, and hence, since F/S_U is separated, $d^m(\pi_U \circ f)(\xi) x^m = 0$ for every $m \in \mathbb{N}$, $x \in E$. Therefore $\pi_U[d^m f(\xi) x^m] = 0$, which implies that $d^m f(\xi) x^m \in S_U$ for every $m \in \mathbb{N}$, $x \in E$. Hence $T_\xi \subset S_U$. Now consider the canonical mapping $\omega_\xi : F \to F/T_\xi$. We have that $\omega_\xi \circ f \in \mathcal{H}(U;F/T_\xi)$, and $d^m(\omega_\xi \circ f)(\xi) x^m = \omega_\xi[d^m f(\xi) x^m] = 0$ for every $m \in \mathbb{N}$, $x \in E$. Hence, by Proposition 31.1, $\omega_\xi \circ f$ is identically zero in U, which implies that $f(U) \subset \subset T_\xi$. Therefore $S_U \subset T_\xi$. Q.E.D.

REMARK 31.1: We have used Proposition 31.1 to prove Proposition 31.2. It is easy to see that Proposition 31.1 follows from Proposition 31.2.

REMARK 31.2: When applying part b) of Proposition 31.2, one often uses the fact that, for each $\xi \in U$, the closed subspace of F generated by $d^m f(\xi) x^m$, $m \in \mathbb{N}$, $x \in E$, coincides with the closed subspace generated by $f(\xi)$ and $d^m f(\xi)(x_1,\ldots,x_m)$, $m \in \mathbb{N}^*$, $x_1,\ldots,x_m \in E$. This can be proved easily using the polarization formula.

CHAPTER 33

THE MAXIMUM SEMINORM THEOREM

PROPOSITION 33.1 Let F be seminormed, $f \in \mathcal{H}(U;F)$ and $\xi \in U$. If X is a non-empty subset of U with the property that $t \in X$ implies $(1-\lambda)\xi+\lambda t \in X$ for $\lambda \in \mathbb{C}$, $|\lambda| = 1$, and $(1-\lambda)\xi+\lambda t \in U$ for $\lambda \in \mathbb{C}$, $|\lambda| \leq 1$, then

$$\|f(\xi)\| \leq \sup\{\|f(x)\| : x \in X\}.$$

PROOF: Let $t \in X$. Applying Proposition 25.2 with x replaced by $t-\xi$, we have

$$f(\xi) = \frac{1}{2\pi i} \int_{|\lambda|=1} \frac{f[(1-\lambda)\xi+\lambda t]}{\lambda} d\lambda$$

and hence

$$\|f(\xi)\| \leq \sup_{\substack{|\lambda|=1 \\ t\in X}} \|f[(1-\lambda)\xi+\lambda t]\| \leq \sup\{\|f(x)\| : x \in X\},$$

where $\|\cdot\|$ denotes the seminorm considered in F. Q.E.D.

COROLLARY 33.1 Let F be seminormed, $f \in \mathcal{H}(U)$ and $\xi \in U$. Let V be a closed ξ-balanced neighbourhood of ξ in U such that the boundary ∂V of V in U is non-empty. Then

$$\|f(\xi)\| \leq \sup\{\|f(x)\| : x \in \partial V\}.$$

PROPOSITION 33.2 Let F be seminormed, U conncected and $f \in \mathcal{H}(U;F)$. If the mapping $\|f\|: x \in U \longmapsto \|f(x)\| \in \mathbb{R}$ has a local maximum at a point ξ of U, then $\|f\|$ is constant in a neighbourhood of ξ and $\|f(\xi)\|$ is the minimum of $\|f\|$ in U.

PROOF: When $E = F = \mathbb{C}$ this proposition is simply the classical Maximum Modulus Theorem, which states that if $|f| : x \in U \to |f(x)| \in \mathbb{R}$ has a local maximum in U, then f is constant in U.

Now let E be arbitrary, and let $F = \mathbb{C}$. We claim that if $|f| : x \in U \mapsto |f(x)| \in \mathbb{R}$ has a local maximum at $\xi \in U$, then f is constant in U. To see this, let U_ξ be the largest ξ-balanced subset of U, and let $x \in U_\xi$. Then $V = \{\lambda \in \mathbb{C} : (1-\lambda)\xi + \lambda x \in U\}$ is an open subset of $\mathbb{C}$, and $\bar{B}_1(0) \subset V$. Define $g \in \mathcal{H}(V;\mathbb{C})$ by $g(\lambda) = f[(1-\lambda)\xi + \lambda x]$ for $\lambda \in V$. Since $|g|$ has a local maximum at 0, it follows that g is constant in the connected component of V which contains 0. In particular, g is constant in $\bar{B}_1(0)$, and hence $g(1) = g(0)$. Therefore $f(x) = f(\xi)$, and so f is constant in U_ξ. Since U_ξ is a neighbourhood of ξ in U, it follows from Proposition 31.1 that f is constant in U.

Finally, let us consider the general case, where E and F are arbitrary. By the Hahn-Banach Theorem, there exists $\psi \in F'$ such that $\|\psi\| \leq 1$ and $\psi[f(\xi)] = \|f(\xi)\|$. Let V be a neighbourhood of ξ in U such that $\|f(\xi)\| \geq \|f(x)\|$ for every $x \in V$. Then

$$|\psi[f(x)]| \leq \|f(x)\| \leq \|f(\xi)\| = \psi[f(\xi)]$$

for every $x \in V$. Thus $\psi \circ f \in \mathcal{H}(U)$ and $|\psi \circ f|$ has a local maximum at ξ. From the cases already considered, it follows that $\psi \circ f$ is constant in U. Therefore, if $x \in U$,

$$\|f(\xi)\| = \psi[f(\xi)] = \psi[f(x)] \leq \|f(x)\|.$$

Hence $\|f(\xi)\| = \|f(x)\|$ for every $x \in V$, and $\|f(\xi)\| \leq \leq \|f(x)\|$ for every $x \in U$. Q.E.D.

COROLLARY 33.2 Under the conditions of the Proposition, if $\|f\|$ has a maximum at a point of U, then $\|f\|$ is constant in U.

EXAMPLE 33.1 Let $E = \mathbb{C}$, and let $F = \mathbb{C}^2$ with the supremum norm. Then $f: z \in E \to (1,z) \in F$ is holomorphic in E, and for $|z| \leq 1$ we have $\|f(z)\| = \sup\{1,|z|\} = 1$. However, for $|z| > 1$, $\|f(z)\| = |z|$. Thus $\|f\|$ is constant in a neighborhood of the origin, but $\|f\|$ is not constant in E; $\|f(0)\|$ is the minimum of $\|f\|$ in E.

Example 33.1 shows that, in contrast to the case $F = \mathbb{C}$, we cannot, in general, conclude in Proposition 30.2 that f is constant in a neighbourhood V of ξ, but only that $\|f\|$ is constant in V. This conclusion is valid if we impose another condition on F:

PROPOSITION 33.3 Let F be a normed space, let U be connected, and let $f \in \mathcal{H}(U;F)$. Suppose that F has the following property: if $\eta \in F$, $\|\eta\| = 1$, there exists $\psi \in F'$ such that $\|\psi\| = 1$, $\psi(\eta) = 1$ and $|\psi(y)| < 1$ for every $y \in F$, $y \neq \eta$, $\|y\| \leq 1$. Then, if $\|f\|: x \in U \mapsto \|f(x)\| \in \mathbb{R}$ has a local maximum at a point ξ in U, f is constant in U.

PROOF: If $\|f(\xi)\| = 0$, the result is trivial. Suppose that

$\|f(\xi)\| > 0$. Multiplying f by a suitable constant, we may assume that $\|f(\xi)\| = 1$. Let $\psi \in F'$ be such that $\|\psi\| = 1$, $\psi[f(\xi)] = 1$, and $|\psi(y)| < 1$ when $y \in F$, $\|y\| \leq 1$, $y \neq f(\xi)$. Let V be a neighbourhood of ξ in U such that $\|f(x)\| \leq \|f(\xi)\| = 1$ for every $x \in V$. Then $\psi \circ f \in \mathcal{H}(U)$ and, for every $x \in V$,

$$|\psi[f(x)]| \leq \|f(x)\| \leq 1 = \psi[f(\xi)].$$

Therefore $|\psi \circ f|$ has a local maximum in U, and it follows from the proof of Proposition 33.2 that $\psi \circ f$ is constant in U. Thus for $x \in V$ we have $\psi[f(x)] = \psi[f(\xi)] = 1$, while $\|f(x)\| \leq 1$, which implies that $f(x) = f(\xi)$. Therefore f is constant in V, and so, by uniqueness of holomorphic continuation, f must be constant in U. Q.E.D.

REMARK 33.1: This proposition applies when F is a Hilbert space, and in particular when $F = \mathbb{C}$.

CHAPTER 34

PROJECTIVE AND INDUCTIVE LIMITS AND HOLOMORPHY

PROPOSITION 34.1 Let $\{F_i\}_{i \in I}$ be a family of complex locally convex spaces, let F be a complex vector space, and for each $i \in I$, let $\rho_i: F \to F_i$ be a linear mapping. Let F be given the projective limit topology defined by the mappings ρ_i, $i \in I$. If U is a non-empty open subset of a complex locally convex space E, and f is a mapping of U into F, then $f \in H(U;F)$ if and only if $\rho_i \circ f \in H(U;F_i)$ for every $i \in I$.

PROOF: If $f \in H(U;F)$, it is obvious that $\rho_i \circ f \in H(U;F_i)$. Conversely, suppose that $\rho_i \circ f \in H(U;F_i)$. For each $i \in I$, let $\hat{F}_i$ be a completion of F_i. If $G = \prod_{i \in I} F_i$ and $\hat{G} = \prod_{i \in I} \hat{F}_i$, then $\hat{G}$ is a completion of G. Define $g: U \to G$ by

$$g: x \in U \longmapsto (\rho_i \circ f(x))_{i \in I} \in G.$$

Then $g \in H(U;G)$. Now define $\rho: F \to G$ by

$$\rho: y \in F \longmapsto (\rho_i(y))_{i \in I} \in G,$$

and let $S = \rho(F)$. Then $g(U) \subset S$, and hence $g \in H(U;S)$. Since the topology of F is the inverse image of the topology of S under the mapping ρ, and $g = \rho \circ f$, it follows that $f \in H(U;F)$. Q.E.D.

REMARK 34.1: In the proof of Proposition 34.1 we have made use of the following fact: if $\rho: F \to G$ is linear and surjective, and the topology of F is the inverse image under ρ of the topology of G, then $\rho \circ f \in H(U;F)$ if and only if $f \in H(U;F)$.

COROLLARY 34.1 If F is a complex locally convex space, then $H(U;F) = \bigcap_{\beta \in CS(F)} H(U;F_\beta)$.

COROLLARY 34.2 If F is a complex locally convex space, and if WF denotes the space F with the weak topology $\sigma(F,F')$, then

$$H(U;WF) = \{f: U \to F : \psi \circ f \in \mathcal{H}(U) \text{ for every } \psi \in F'\}.$$

Corollary 34.1 follows immediately from Proposition 34.1, and Corollary 34.2 follows from Corollary 34.1.

Our next example shows that it is not, in general, possible to replace the symbol H by $\mathcal{H}$ in Corollary 34.2. Thus Proposition 34.1 and Corollary 34.1 are not, in general, valid if H is replaced by $\mathcal{H}$.

EXAMPLE 34.1 Suppose that F is not differentially stable. Then, by Proposition 29.2, WF is not differentially stable. It follows that there exists a non-empty open subset U of $\mathbb{C}$ such that $H(U;WF) \neq \mathcal{H}(U;WF)$. Hence Corollary 34.2 must be false for $\mathcal{H}(U;WF)$.

PROPOSITION 34.2 Let $\{E_m\}_{m \in \mathbb{N}}$ be a sequence of complex locally convex spaces, let E be a complex vector space, for each $m \in \mathbb{N}$ let $\rho_m: E_m \to E$ be a linear mapping, and let $\sigma_m: E_m \to E_{m+1}$ be a compact linear mapping such that $\rho_m = \rho_{m+1} \circ \sigma_m$ for every $m \in \mathbb{N}$. Suppose that $E = \bigcup_{m \in \mathbb{N}} \rho_m(E_m)$, and let E be given the locally convex inductive limit topology defined by the mappings ρ_m, $m \in \mathbb{N}$. If U is an open subset of E, let $U_m = \rho_m^{-1}(U)$, $m \in \mathbb{N}$, and suppose that U_o is non-empty. If F is a complex locally convex space, and f a mapping of U into F, then $f \in \mathcal{H}(U;F)$ if and only if $f \circ \rho_m \in \mathcal{H}(U_m;F)$ for every $m \in \mathbb{N}$.

REMARK 34.2: To say that the linear mapping $\sigma_m: E_m \to E_{m+1}$ is compact means that there exists a neighbourhood V_m of the origin in E_m such that $\sigma_m(V_m)$ is relatively compact in E_{m+1}. The compatibility condition in the statement of the proposition concerning the mappings σ_m and ρ_m states that the diagram

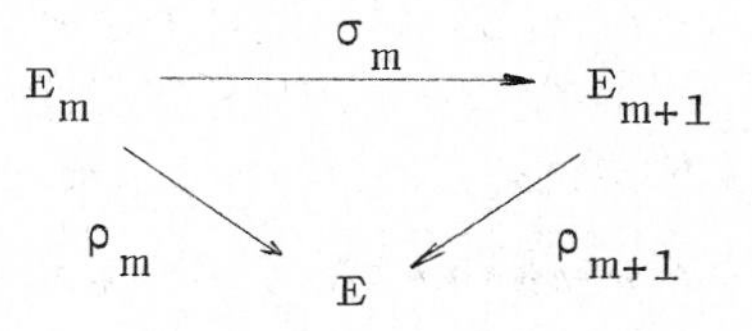

commutes for every $m \in \mathbb{N}$. The condition $U_o = \rho_o^{-1}(U) \neq \emptyset$ implies that $U_m = \rho_m^{-1}(U) \neq \emptyset$ for every $m \in \mathbb{N}$. This follows by induction from the relation:

$$(*) \qquad \sigma_m[\rho_m^{-1}(U)] \subset \rho_{m+1}^{-1}(U).$$

To prove (*), let $x_m = \rho^{-1}(x)$, where $x \in U$. Then $\rho_m(x_m) = x \in U$, which implies that $\rho_{m+1}[\sigma_m(x_m)] = \rho_m(x_m) = x \in U$, and hence $\sigma_m(x_m) \in \rho_{m+1}^{-1}(U)$.

PROOF OF PROPOSITION 34.2: If $f \in \mathcal{H}(U;F)$, it is clear that $f\circ\rho_m \in \mathcal{H}(U_m;F)$ for every $m \in \mathbb{N}$. Conversely, suppose that $f\circ\rho_m \in \mathcal{H}(U_m;F)$ for every $m \in \mathbb{N}$. We show first that f is finitely holomorphic.

Suppose that E is separated, and let S be a finite dimensional subspace of E. We claim that there exists $m \in N$ for which E_m contains a subspace S_m such that ρ_m is a topological isomorphism of S_m with S. Let $\{y_1,\ldots,y_k\}$ be a basis for S. For each $i = 1,\ldots,k$ there exists $n_i \in \mathbb{N}$ and $x_{n_i} \in E_{n_i}$ such that $y_i = \rho_{n_i}(x_{n_i})$. Choose $m \in \mathbb{N}$ greater than the maximum of $\{n_1,\ldots,n_k\}$. Then

$$y_1 = \rho_{n_1}(x_{n_1}) = \rho_{n_1+1}\circ\sigma_{n_1}(x_{n_1})=\ldots=\rho_m\circ\sigma_{m-1}\circ\sigma_{m-2}\circ\ldots\circ\sigma_{n_1}(x_{n_1}),$$

and similarly, for each $i = 1,\ldots,k$ we have

$$y_i = \rho_m\circ\sigma_{m-1}\circ\sigma_{m-2}\circ\ldots\circ\sigma_{n_i}(x_{n_i}).$$

Let $z_i = \sigma_{m-1}\circ\sigma_{m-2}\circ\ldots\circ\sigma_{n_i}(x_{n_i})$, $i = 1,\ldots,k$. Then $z_i \in E_m$ for every i, and $\{z_1,\ldots,z_k\}$ is linearly independent, since $\{y_1,\ldots,y_k\}$ is linearly independent, and $y_i = \rho_m(z_i)$. Hence, if S_m denotes the subspace of E_m generated by $\{z_1,\ldots,z_k\}$, then ρ_m is an isomorphism of the vector spaces S_m and S. Furthermore, ρ_m is continuous and S is separated. Therefore ρ_m is also a homeomorphism of S_m with S.

Furthermore, we have

$$\rho_m(U_m\cap S_m) = \rho_m[\rho_m^{-1}(U) \cap \rho_m^{-1}(S)] = \rho_m[\rho_m^{-1}(U\cap S)] = U \cap S,$$

and so $\rho_m(U_m\cap S_m) \neq \emptyset$ if $U \cap S \neq \emptyset$. Now, since

$f \circ \rho_m \in \mathcal{H}(U_m;F)$, we have that $(f \circ \rho_m)/U_m \cap S_m \in \mathcal{H}(U_m \cap S_m;F)$. But $(f \circ \rho_m)/U_m \cap S_m = (f/U \cap S) \circ (\rho_m/U_m \cap S_m)$, and $\rho_m/U_m \cap S_m : U_m \cap S_m \to U \cap S$ is bijective. Therefore

$$f/U \cap S = [(f \circ \rho_m)/U_m \cap S_m] \circ [(\rho_m/U_m \cap S_m)^{-1}/U \cap S],$$

and it follows that f is holomorphic in $U \cap S$. Hence f is finitely holomorphic.

We complete the proof by showing that f is amply bounded in U. For this, it suffices to consider the case where F is semnormed. It is not difficult to show that we may assume $0 \in U$, and that it suffices to prove that f is locally bounded at 0.

We show first that $\sigma_m : E_m \to E_{m+1}$ is continuous for every $m \in N$. To see this, let V_m be a neighbourhood of the origin in E_m such that $\sigma_m(V_m)$ is relatively compact in E_{m+1}. Then $\sigma_m(V_m)$ is bounded, and so, if W_{m+1} is a neighbourhood of the origin in E_{m+1}, there exists $\lambda > 0$ such that

$$\lambda\sigma_m(V_m) \subset V_{m+1} \Leftrightarrow \sigma_m^{-1}\sigma_m(\lambda V_m) \subset \sigma_m^{-1}(W_{m+1})$$

$$\Rightarrow \lambda V_m \subset \sigma_m^{-1}(V_{m+1});$$

that is, $\sigma_m^{-1}(W_m)$ is a neighbourhood of the origin in E_m. Therefore σ_m is continuous.

We now assume $0 \in U$, and prove that f is locally bounded at 0. Let $U' \subset U$ be a closed neighbourhood of the origin in E, and consider for each $m \in \mathbb{N}$ the set $U'_m = \rho_m^{-1}(U')$; each U'_m is a closed neighbourhood of the origin in E_m, and $U'_m \subset U_m = \rho_m^{-1}(U)$. Also,

$$\sigma_m(U'_m) = \sigma_m(\rho_m^{-1}(U')) = \sigma_m(\sigma_m^{-1}\circ\rho_{m+1}^{-1}(U')) \subset \rho_{m+1}^{-1}(U') = U'_{m+1}.$$

Let $f_m = f\circ\rho_m$, $m \in \mathbb{N}$. Since f_o is locally bounded at zero, there is a neighbourhood V'_o of zero contained in U'_o, and real numbers M_o and M such that

$$\sup\{\|f_o(x)\| : x \in V'_o\} = M_o < M.$$

Since σ_o is compact, there exists a convex neighbourhood V''_o of the origin in E_o such that $\overline{\sigma_o(V''_o)}$ is compact in E_1. Let $V_o = V'_o \cap V''_o$. Then

$$\sup\{\|f_o(x)\| : x \in V_o\} \leq M_o < M,$$

and $\sigma_o(V_o) \subset \sigma_o(U'_o) \subset U'_1$, which is closed in E_1; hence $\overline{\sigma_o(V_o)} \subset \overline{\sigma_o(U'_o)} \subset U'_1 \subset U_1 = \rho_1^{-1}(U)$. Also, since $V_o \subset V''_o$, $\overline{\sigma_o(V_o)} \subset \overline{\sigma_o(V''_o)}$, and so $\overline{\sigma_o(V_o)}$ is compact in E_1.

Now suppose that for $m \in \mathbb{N}$ we have a convex neighbourhood V_m of the origin in E_m such that $\overline{\sigma_m(V_m)}$ is compact in E_{m+1}, $\overline{\sigma_m(V_m)} \subset U'_{m+1} = \rho_{m+1}^{-1}(U')$, and

$$\sup\{\|f_m(x)\| : x \in V_m\} = M_m < M.$$

Since $f_m = f_{m+1}\circ\sigma_m$, we then have

$$\sup\{\|f_{m+1}(y)\| : y \in \sigma_m(V_m)\} =$$

$$= \sup\{\|f_m(x)\| : x \in V_m\} = M_m < M,$$

and hence

$$\sup\{\|f_{m+1}(y)\| : y \in \overline{\sigma_m(V_m)}\} = M_m.$$

Choose $\varepsilon > 0$ such that $M_{m+1} = M_m + \varepsilon < M$. Since σ_{m+1} is compact, there exists a convex neighbourhood $V'_{m+1}(0)$ of the origin in E_{m+1} such that $\overline{\sigma_{m+1}(V'_{m+1}(0))}$ is compact in E_{m+2}. Since f_{m+1} is uniformly continuous on the compact set $\overline{\sigma_m(V_m)}$, there exists a closed neighbourhood $V''_{m+1}(0)$ of the origin in E_{m+1} such that

$$\sup\{\|f_{m+1}(z)\| : z \in \bigcup_{y \in \overline{\sigma_m(V_m)}} (y + V''_{m+1}(0))\} \leq M_m + \varepsilon =$$

$$= M_{m+1} < M,$$

and $\overline{\sigma_m(V_m)} \subset \bigcup_{y \in \overline{\sigma_m(V_m)}} (y + \{V''_{m+1}(0)^{\circ}) \subset U'_{m+1}$, where $\{V''_{m+1}(0)\}^{\circ}$ is the interior of $V''_{m+1}(0)$ in E_{m+1}.

Let $V_{m+1}(0) = \{V'_{m+1}(0) \cap V''_{m+1}(0)\}^{\circ}$. Then there exist $y_1, \ldots, y_h \in \overline{\sigma_m(V_m)}$ such that

$$\overline{\sigma_m(V_m)} \subset \bigcup_{k=1}^{h} (y_k + V_{m+1}(0)) \subset U'_{m+1}.$$

Let $V'_{m+1} = \bigcup_{k=1}^{h} (y_k + V_{m+1}(0))$. Then:

a) $\sigma_{m+1}(V'_{m+1}) = \bigcup_{k=1}^{h} \{\sigma_{m+1}(y_k) + \sigma_{m+1}(V_{m+1}(0))\} \subset$
$\subset \sigma_{m+1}(U'_{m+1}) \subset U'_{m+2}$, which implies that $\overline{\sigma_{m+1}(V'_{m+1})} \subset \overline{U'_{m+2}} =$
$= U'_{m+2} \subset U_{m+2}$, and

b) $\sup\{\|f_{m+1}(z)\| : z \in V'_{m+1}\} \leq M_{m+1} < M$.

Now, since $\overline{\sigma_m(V_m)}$ is compact, and is contained in V'_{m+1}, there exists $\alpha \in CS(E_{m+1})$ such that

$$\overline{\sigma_m(V_m)} \subset B_{\alpha,1}[\sigma_m(V_m)] \subset V'_{m+1},$$

where $B_{\alpha,1}[\sigma_m(V_m)] = \bigcup_{z \in \sigma_m(V_m)} (z + B_{\alpha+1}(0))$.

Therefore $\sigma_{m+1}[B_{\alpha,1}(\overline{\sigma_m(V_m)})] \subset \sigma_{m+1}(V'_{m+1})$, and hence $\sigma_{m+1}[B_{\alpha,1}(\overline{\sigma_m(V_m)})]$ is relatively compact. We define $V_{m+1} = \sigma_{m+1}[B_{\alpha,1}(\overline{\sigma_m(V_m)})]$. Then

$$V_{m+1} \subset V'_{m+1} \subset U'_{m+1} \subset \rho^{-1}_{m+1}(U),$$

and so $\rho_{m+1}(V_{m+1}) \subset U$. Also, since $\sigma_m(V_m) \subset V_{m+1}$, we have

$$\rho_{m+1} \circ \sigma_m(V_m) = \rho_m(V_m) \subset \rho_{m+1}(V_{m+1}).$$

Let $V = \bigcup_{m \in \mathbb{N}} \rho_m(V_m)$. Then V is a neighbourhood of the origin in U, and $\sup\{\|f(x)\| : x \in V\} < M$. Therefore f is locally bounded at the origin. Q.E.D.

DEFINITION 34.1 A locally convex space E is a Silva space if there exists a sequence $\{E_m\}_{m \in \mathbb{N}}$ of Banach spaces such that E_m is a vector subspace of E, $E_m \subset E_{m+1}$, the inclusion mapping $E_m \to E_{m+1}$ is compact for every $m \in \mathbb{N}$, $E = \bigcup_{m \in \mathbb{N}} E_m$, and the topology of E is the locally convex inductive limit topology corresponding to this decomposition.

REMARK 34.3: Silva spaces, apart from being of interest in themselves, appear frequently in holomorphy, even in finite dimensions. For example, if K is a compact subset of $\mathbb{C}^n$, $n \in \mathbb{N}^*$, the space $\mathcal{H}(K)$ of holomorphic germs on K with its natural topology, is a Silva space. If U is a non-empty open subset of $\mathbb{R}^n$, the space $\mathcal{E}'(U)$ of distributions on U with compact support is a Silva space. If U is a non-empty open subset of $\mathbb{C}^n$, $\mathcal{H}'(U) = [\mathcal{H}(U)]'$ is a Silva space.

We point out that the hypotheses of Proposition 34.1 characterise the class of Silva spaces, by virtue of the following result: If E is a vector space, $\{E_m\}_{m\in\mathbb{N}}$ a sequence of locally convex spaces, $\rho_m: E_m \to E$ a linear mapping and $\sigma_m: E_m \to E_{m+1}$ a compact linear mapping, with $\rho_m = \rho_{m+1}\circ\sigma_m$ for every $m \in \mathbb{N}$, if $E = \bigcup_{m\in\mathbb{N}} \rho_m(E_m)$, and if E carries the corresponding locally convex inductive limit topology, then E is a Silva space.

REMARK 34.4: Proposition 21.1 is not, in general, valid if the assumption that the mappings σ_m are compact is dropped, or if the sequence $\{E_m\}_{m\in\mathbb{N}}$ is replaced by an uncountable family of spaces.

CHAPTER 35

TOPOLOGIES ON $\mathcal{H}(U;F)$

DEFINITION 35.1 Let p be a seminorm on $\mathcal{H}(U;F)$, and let K be a compact subset of U. p is said to be ported by K if there exists $\beta \in CS(F)$ such that for every open subset V of U with $K \subset V$ there exists a real number $C(V) > 0$, with the property that

$$p(f) \leq C(V) \sup_{x \in V} \beta\{f(x)\} \quad \text{for every} \quad f \in \mathcal{H}(U;F).$$

The Nachbin topology, denoted by τ_ω, is the locally convex topology on $\mathcal{H}(U;F)$ defined by the seminorms on $\mathcal{H}(U;F)$ which are ported by compact subsets of U.

EXAMPLE 35.1 Let $K \subset U$ be compact and $\beta \in CS(F)$. Then the seminorm

$$f \in \mathcal{H}(U;F) \mapsto p(f) = \sup_{x \in K} \beta\{f(x)\}$$

is ported by K.

REMARK 35.1 As K ranges over the compact subsets of U, and β over $CS(F)$, the seminorms $f \in \mathcal{H}(U;F) \mapsto \sup_{x \in K} \beta\{f(x)\}$ define the compact-open topology, τ_o, on $\mathcal{H}(U;F)$. Thus, it follows from Example 35.1 that $\tau_o \leq \tau_\omega$ on $\mathcal{H}(U;F)$.

EXAMPLE 35.2 Suppose that F is separated. Let K be a

compact subset of U, $\beta \in CS(F)$, and let $\{c_m\}_{m\in\mathbb{N}}$ be a sequence of non-negative real numbers such that $c_m^{1/m} \to 0$ as $m \to \infty$. Let L be a bounded subset of E. Then the mapping

$$f \in \mathcal{H}(U;F) \mapsto p(f) = \sum_{m=0}^{\infty} c_m \sup_{\substack{x\in K \\ t\in L}} \beta\{\frac{1}{m!}\hat{d}^m f(x)(t)\} \in \mathbb{R}$$

defines a seminorm which is ported by K. To see this, let $f \in \mathcal{H}(U;F)$. Then there exists $\alpha \in CS(E)$ such that $V = B_{\alpha,1}(K) = \bigcup_{x\in K} B_{\alpha,1}(x) \subset U$ and $\sup_{x\in V} \beta\{f(x)\} = M < \infty$. Choose $\rho > 0$ such that $x+\lambda t \in V$ for $t \in L$, $x \in K$, $\lambda \in \mathbb{C}$, $|\lambda| \le \rho$. Then, by the Cauchy integral formula,

$$\frac{1}{m!}\hat{d}^m f(x)(t) = \frac{1}{2\pi i}\int_{|\lambda|=\rho} \frac{f(x+\lambda t)}{\lambda^{m+1}}\, d\lambda$$

and hence

$$\beta\{\frac{1}{m!}\hat{d}^m f(x)(t)\} \le \frac{M}{\rho^m} \quad \text{for } x \in K, \ t \in M, \ m \in \mathbb{N}.$$

Therefore

$$c_m \sup_{\substack{x\in K \\ t\in L}} \beta\{\frac{1}{m!}\hat{d}^m f(x)(t)\} \le M\left(\frac{c_m^{1/m}}{\rho}\right)^m$$

for every $m \in \mathbb{N}$. For m sufficiently large, $c_m^{1/m}/\rho \le 1/2$, and hence the series which defines $p(f)$ converges. It follows easily that p is a seminorm on $\mathcal{H}(U;F)$. We now show that p is ported by K. Let V be an open subset of E, with $K \subset V \subset U$. Then there exists $\alpha \in CS(E)$ such that $B_{\alpha,1}(K) \subset V$, and there exists $r > 0$, depending only on V such that $x+\lambda t \in \bar{B}_{\alpha,1}(K)$ for every $x \in K$, $t \in L$ and $\lambda \in \mathbb{C}$, $|\lambda| \le r$. Therefore

$$\sum_{m=0}^{\infty} c_m \sup_{\substack{x\in K\\ t\in L}} \beta\{\frac{1}{m!}\hat{d}^m f(x)(t)\} \le \sup_{y\in V} \beta\{f(y)\} \sum_{m=0}^{\infty} \left(\frac{c_m^{1/m}}{r}\right)^m$$

for every $f \in \mathcal{H}(U;F)$. Since $\sum_{m=1}^{\infty} \left(\frac{c_m^{1/m}}{r}\right)^m$ is independent of f, we may define

$$C(V) = \sum_{m=0}^{\infty} \left(\frac{c_m^{1/m}}{r}\right)^m .$$

Therefore $p(f) \le C(V) \sup_{y\in V} \beta\{f(y)\}$ for every $f \in \mathcal{H}(U;F)$, and so p is ported by K.

PROPOSITION 35.1 Let F be separated, let $\xi \in U$, and let K be a compact ξ-balanced subset of U. A seminorm p on $\mathcal{H}(U;F)$ is ported by K if and only if there exists $\beta \in CS(F)$ such that for every open subset V of U containing K, there exists $C(V) > 0$ with the property that

$$p(f) \le C(V) \sum_{m=0}^{\infty} \sup_{x\in V} \beta\{\frac{1}{m!}\hat{d}^m f(\xi)(x-\xi)\}$$

for every $f \in \mathcal{H}(U;F)$.

PROOF: To prove that this condition is necessary, let p be a seminorm on $\mathcal{H}(U;F)$ which is ported by K, and let $\beta \in CS(F)$ be a seminorm associated with p by Definition 35.1. If V is an open subset of U containing K, let V_ξ be the largest ξ-balanced subset of V. Then V_ξ is open and contains K, and hence there exists $C(V_\xi) > 0$ such that

a) $p(f) \le C(V_\xi) \sup_{x\in V_\xi} \beta\{f(x)\}$ for every $f \in \mathcal{H}(U;F)$.

By Proposition 27.2, the Taylor series of f at ξ converges to f in V_ξ. Hence, for every $f \in \mathcal{H}(U;F)$ and every $x \in V_\xi$ we have

$$\beta\{f(x)\} \leq \sum_{m=0}^{\infty} \beta\{\frac{1}{m!} \hat{d}^m f(\xi)(x-\xi)\}.$$

Taking this in conjunction with a), we have

$$\text{b)} \quad p(f) \leq C(V_\xi) \sum_{m=0}^{\infty} \sup_{x \in V} \beta\{\frac{1}{m!} \hat{d}^m f(\xi)(x-\xi)\}$$

for every $f \in \mathcal{H}(U;F)$. Since V_ξ depends only on V, this proves the necessity of the given condition.

To prove that this is sufficient, suppose that there exists $\beta \in CS(F)$ such that for every open subset V of U containing K there is a constant $C(V) > 0$ such that

$$p(f) \leq C(V) \sum_{m=0}^{\infty} \sup_{x \in V} \beta\{\frac{1}{m!} \hat{d}^m f(\xi)(x-\xi)\}$$

for every $f \in \mathcal{H}(U;F)$. Since K is compact and ξ-balanced, there exists an open set W, $K \subset W \subset V$, and $\rho > 1$, such that $(1-\lambda)\xi + \lambda x \in V$ if $x \in W$ and $\lambda \in \mathbb{C}$, $|\lambda| \leq \rho$. There exists $C(W) > 0$ such that

$$p(f) \leq C(W) \sum_{m=0}^{\infty} \sup_{x \in W} \beta\{\frac{1}{m!} \hat{d}^m f(\xi)(x-\xi)\}$$

for every $f \in \mathcal{H}(U;F)$. By the Cauchy integral formula (Proposition 25.2) it follows that

$$p(f) \leq C(W) \sum_{m=0}^{\infty} \frac{1}{\rho^m} \sup_{t \in V} \beta\{f(t)\},$$

and hence

$$p(f) \leq \frac{\rho}{\rho-1} C(W) \sup_{t \in V} \beta\{f(t)\}$$

for every $f \in \mathcal{H}(U;F)$. Since $\frac{\rho}{\rho-1} C(W)$ is independent of f, this shows that p is ported by K. Q.E.D.

PROPOSITION 35.2 Let $\{f_\lambda\}_{\lambda \in \Lambda}$ be a net in $\mathcal{H}(U;F)$. Suppose that for every $\beta \in CS(F)$ and every compact subset K of U there exists an open subset V of U containing K such that

$$\lim_{\lambda} \sup_{x \in V} \beta\{f(x)-f_\lambda(x)\} = 0$$

for some $f \in \mathcal{H}(U;F)$. Then $\lim_{\lambda} f_\lambda = f$ in the topology τ_ω.

PROOF: Let p be a τ_ω-continuous seminorm on $\mathcal{H}(U)$. Then p is ported by some compact set $K \subset U$. Let $\beta \in CS(F)$ be a seminorm corresponding to p in Definition 35.2, so that for every open subset W of U containing K, there exists $C(W) > 0$ for which

a) $p(f) \leq C(W) \sup_{x \in W} \beta\{f(x)\}$ for every $f \in \mathcal{H}(U;F)$.

By hypothesis, there exists an open subset V of U containing K such that for every $\varepsilon > 0$ there exists $\lambda_o \in \Lambda$, with

b) $\sup_{x \in V} \beta\{f(x)-f_\lambda(x)\} \leq \varepsilon$ for $\lambda \geq \lambda_o$.

Combining a) and b), we find that $p(f-f_\lambda) \leq \varepsilon C(V)$ when $\lambda \geq \lambda_o$. Therefore $\{f_\lambda\}_{\lambda \in \Lambda}$ converges to f in the topology τ_ω. Q.E.D.

PROPOSITION 35.3 Let $\xi \in U$, and suppose that U is ξ-balanced. Then for every $f \in \mathcal{H}(U;F)$ the Taylor series of f at ξ converges to f in the topology τ_ω.

PROOF: Since U is ξ-balanced, $U_\xi = U$, and hence, by Proposition 27.1, for every $\beta \in CS(F)$ and every compact subset K of U, there exists an open subset V of U containing K such that

$$\lim_{m\to\infty} \sup_{x\in V} \{f(x)-\tau_{m,f,\xi}(x)\} = 0.$$

Therefore, by Proposition 35.2, the Taylor series of f at ξ converges to f in the topology τ_ω. Q.E.D.

PROPOSITION 35.4 Let F be separated, $f \in \mathcal{H}(U;F)$, and $m \in \mathbb{N}$. Then $\hat{d}^m f \in \mathcal{H}(U; \mathcal{P}_i(^mE;F))$, where $\mathcal{P}_i(^mE;F)$ denotes $\mathcal{P}(^mE;F)$ with the limit topology (see Definition 16.2). Furthermore, if

$$\sum_{k=0}^{\infty} P_k(x-\xi)$$

is the Taylor series of f at ξ, then $\sum_{k=0}^{\infty} \hat{d}^m P_{k+m}(x-\xi)$ is the Taylor series of $\hat{d}^m f$ at ξ.

PROOF: We shall prove the proposition for the bounded topology on $\mathcal{P}(^mE;F)$ (see Definition 16.3). Given $\beta \in CS(F)$, choose $\alpha \in CS(E)$ such that $B_{\alpha,1}(\xi) \subset U$ and

$$\beta\{f(x) - \sum_{k=0}^{h} P_k(x-\xi)\} \to 0$$

uniformly in $B_{\alpha,1}(\xi)$ when $h \to \infty$. Let

$$T_\xi : x \in E \mapsto x-\xi \in E, \quad \text{and} \quad Q_k = P_k \circ T_\xi, \qquad k \in \mathbb{N}.$$

Then $\lim_{h\to\infty} \beta\{f(x) - \sum_{k=0}^{h} Q_k(x)\} = 0$ uniformly in $B_{\alpha,1}(\xi)$.

Let L be a bounded subset of E, and let $r \geq \max\{1, \sup_{x\in L} \alpha(x)\}$. Then, for $x \in B_{\alpha,1/2}(\xi)$, $|\lambda| \leq 1/(3r)$, and $y \in L$,

$$x+\lambda y \in \bar{B}_{\alpha,1/3}(x) \subset B_{\alpha,1}(\xi).$$

Applying the Cauchy integral formula for $y \in L$ and $x \in \bar{B}_{\alpha,r/3}(\xi)$, we have

$$\hat{d}^m[f - \sum_{k=0}^{h} Q_k](x)(y) = \frac{m!}{2\pi i} \int_{|\lambda|=1/3r} \frac{[f - \sum_{h=0}^{k} Q_k](x+\lambda y)}{\lambda^{m+1}} d\lambda .$$

Therefore

$$\sup_{y \in L} \beta[\{\hat{d}^m f(x) - \sum_{k=0}^{h} \hat{d}^m Q_k(x)\}(y)] \leq$$

$$\leq \frac{m!}{(3r)^m} \sup_{z \in B_{\alpha,1}(\xi)} \beta\{f(z) - \sum_{h=0}^{k} P_k(x-\xi)\}$$

for every $x \in \bar{B}_{\alpha,1/3}(\xi)$. Hence

$$\sup_{y \in L} \beta[\{\hat{d}^m f(x) - \sum_{k=0}^{h} \hat{d}^m P_k(x-\xi)\}(y)] \to 0$$

uniformly in $\bar{B}_{\alpha,1/3}(\xi)$ as $h \to \infty$. Therefore $\hat{d}^m f \in \mathcal{H}(U; \mathcal{P}(^mE;F))$ and

$$\sum_{k=0}^{\infty} \hat{d}^m P_k(x-\xi)$$

is the Taylor series of $\hat{d}^m f$ at ξ, where $\mathcal{P}(^mE;F)$ carries the bounded topology. It follows from Remark 21.1 that this series can be written as

$$\sum_{k=0}^{\infty} \hat{d}^m P_{k+m}(x-\xi).$$

The proof for the limit topology on $\mathcal{P}(^mE;F)$ follows easily from the case we have considered. Q.E.D.

DEFINITION 35.2 Let $\tau_i({}^m E;F)$ denote the limit topology on $\mathcal{P}({}^m E;F)$. τ_ℓ denotes the locally convex topology on $\mathcal{H}(U;F)$ defined by the family of seminorms

$$f \in \mathcal{H}(U;F) \longmapsto \sup_{x\in K} q[\hat{d}^m f(x)]$$

where K ranges over the compact subsets of U, and q ranges over the $\tau_i({}^m E;F)$-continuous seminorms on $\mathcal{P}({}^m E;F)$, and $m \in \mathbb{N}$.

REMARK 35.2 We note that this definition is valid, since $\hat{d}^m f \in \mathcal{H}(U; \mathcal{P}_i({}^m E;F)) \subset C(U; \mathcal{P}_i({}^m E;F))$ for every $m \in N$, where $\mathcal{P}_i({}^m E;F)$ denotes the space $\mathcal{P}({}^m E;F)$ with the topology $\tau_i({}^m E;F)$.

PROPOSITION 35.5 $\tau_\ell \leq \tau_\omega$ on $\mathcal{H}(U;F)$.

PROOF: Let K be a compact subset of U, $m \in \mathbb{N}$ and q a $\tau_i({}^m E;F)$-continuous seminorm on $\mathcal{P}({}^m E;F)$. If V is an open subset of U containing K, choose $\alpha_o \in CS(E)$ such that

$$\bar{B}_{\alpha_o,1}(K) \subset V.$$

Since the seminorm q is continuous for the limit topology, there exists $\beta \in CS(F)$ such that $q = \gamma_\beta \circ \pi_\beta$, where π_β is the canonical linear mapping of $\mathcal{P}({}^m E;F)$ into $\mathcal{P}({}^m E;F_\beta)$, and γ_β is a seminorm on $\mathcal{P}({}^m E;F_\beta)$ which is continuous for the inductive limit topology corresponding to the decomposition

$$\mathcal{P}({}^m E;F_\beta) = \bigcup_{\alpha\in CS(E)} \mathcal{P}({}^m E_\alpha;F_\beta).$$

If $f \in \mathcal{H}(U;F)$ is such that $\hat{d}^m f(x) \in \mathcal{P}(^m E_{\alpha_o};F_\beta)$ for every $x \in K$, then there exists $r > 0$ such that

$$1) \qquad \sup_{x\in K} q[\hat{d}^m f(x)] = \sup_{x\in K} \gamma_\beta \circ \pi_\beta [\hat{d}^m f(x)] =$$

$$= \sup_{x\in K} \gamma_\beta [\hat{d}^m f(x)] \le r \sup_{x\in K} \|\hat{d}^m f(x)\|_{\alpha_o,\beta} .$$

On the other hand, if $f \in \mathcal{H}(U;F)$ is such that $\hat{d}^m f(x) \notin$ $\notin \mathcal{P}(^m E_{\alpha_o};F_\beta)$ for some $x \in K$, then 1) is trivial, since in this case

$$\sup_{x\in K} \|\hat{d}^m f(x)\|_{\alpha_o,\beta} = +\infty .$$

Applying the Cauchy inequality we have

$$\|\frac{1}{m!} \hat{d}^m f(x)\|_{\alpha_o,\beta} \le \frac{1}{1^m} \sup_{\alpha_o(t-\xi)=1} \beta\{f(t)\}$$

$$\le \sup_{y\in V} \beta\{f(y)\},$$

and hence, by 1),

$$\sup_{x\in K} q[\hat{d}^m f(x)] \le rm! \sup_{y\in V} \beta\{f(y)\}$$

for every $f \in \mathcal{H}(U;F)$. Therefore the seminorm

$$f \in \mathcal{H}(U;F) \mapsto \sup_{x\in K} q[\hat{d}^m f(x)]$$

is τ_ω-continuous. Q.E.D.

CHAPTER 36

BOUNDED SUBSETS OF $\mathcal{H}(U;F)$

PROPOSITION 36.1 Let F be separated. In order that a subset χ of $\mathcal{H}(U;F)$ be τ_o-bounded, it is necessary and sufficient that for every $\beta \in CS(F)$, every compact subset K of U, and every bounded subset B of E, there exist real numbers C, $c \geq 0$ such that

$$\sup\{\beta[\frac{1}{m!}\hat{d}^m f(t)(x)] : f \in \chi, t \in K, x \in B\} \leq C \cdot c^m$$

for every $m \in \mathbb{N}$.

PROOF: Taking $m = 0$, we see that this condition is evidently sufficient. Conversely, let χ be τ_o-bounded. Consider first the case where B is compact. Choose $r > 0$ such that

$$L = \{t+\lambda x : t \in K, x \in B, \lambda \in \mathbb{C}, |\lambda| \leq r\} \subset U.$$

By the Cauchy inequalities we have

$$\sup\{\beta[\frac{1}{m!}\hat{d}^m f(t)(x)] : f \in \chi, t \in K, x \in B\} \leq$$

$$\leq \frac{1}{r^m} \sup\{\beta[f(y)] : f \in \chi, y \in L\} = C \cdot c^m$$

where $C = \sup\{\beta[f(y)] : f \in \chi, y \in L\}$ and $c = 1/r$.

We now consider the case of a general bounded set B. We argue by contradiction; thus suppose that, taking $C = k$

and $c = k^2$ for $k = 1,2,\ldots$, there exist $m_k \in N$, $f_k \in \chi$, $t_k \in K$ and $x_k \in L$ such that

$$\beta[\frac{1}{m_k!}\hat{d}^{m_k}f_k(t_k)(x_k)] > k\cdot(k^2)^{m_k} \quad \text{for every } k.$$

Let $y_k = x_k/k$. Then

$$\beta[\frac{1}{m_k!}\hat{d}^{m_k}f(t_k)(y_k)] \geq k\cdot k^{m_k} \quad \text{for } k = 1,2,\ldots .$$

The set $J = \{0\} \cup \{y_1,y_2,\ldots\}$ is compact, since $y_k \to 0$ as $k \to \infty$, and we have

$$\sup\{\beta[\frac{1}{m_k!}\hat{d}^{m_k}f(t)(y)] : f \in \chi,\ t \in K,\ y \in J\} > k\cdot k^{m_k}$$

for $k = 1,2,\ldots$, contradicting the first part of the proof.

Q.E.D.

PROPOSITION 36.2 If E is metrizable then a subset χ of $\mathcal{H}(U;F)$ is τ_ω-bounded if and only if χ is τ_o-bounded.

PROOF: Since $\tau_\omega \geq \tau_o$, every τ_ω-bounded set is τ_o-bounded. Conversely, suppose that χ is τ_o-bounded. Then

$$\sup\{\beta[f(x)] : f \in \chi,\ x \in K\} < +\infty$$

for every $\beta \in CS(F)$, and every compact subset K of U. We claim that if $K \subset U$ is compact, then

(*) for every $\beta \in CS(F)$ there exists an open subset V of U containing K such that

$$\sup\{\beta[f(x)] : f \in \chi,\ x \in V\} < +\infty.$$

Let d be a metric on E which defines the topology of E.

If (*) is false, there exists $\beta \in CS(F)$ such that, for each $n \in \mathbb{N}^*$ there exist $x_n \in U$ and $f_n \in \chi$ for which

$$d(K,x_n) < 1/n \quad \text{and} \quad \beta\{f_n(x_n)\} \geq n.$$

But this is impossible, since $K \cup \{x_n\}_{n\in\mathbb{N}^*}$ is compact. Thus (*) is proved, and it follows that χ is τ_ω-bounded.

Q.E.D.

REMARK 36.1: If E is not metrizable, we can no longer assert that every τ_o-bounded subset of $\mathcal{H}(U;F)$ is τ_ω-bounded (see [9], pages 53-55).

REMARK 36.2: Let us recall that a subset χ of $\mathcal{H}(U;F)$ is said to be amply bounded if, for every $\xi \in U$ and every $\beta \in CS(F)$ there is a open subset V of U containing ξ such that

$$\sup\{\beta[f(x)] : f \in \chi,\ x \in V\} < +\infty.$$

PROPOSITION 36.3 Every amply bounded subset of $\mathcal{H}(U;F)$ is τ_ω-bounded.

PROOF: Let p be a seminorm on $\mathcal{H}(U;F)$ which is ported by a compact set $K \subset U$, and let $\beta \in CS(F)$ be a seminorm associated with p by Definition 22.1. Thus, if V is an open subset of U containing K, there exists $C(V) > 0$ such that

a) $\quad p(f) \leq C(V) \sup\limits_{x\in V} \beta\{f(x)\}$ for every $f \in \mathcal{H}(U;F)$.

Let $\chi \subset \mathcal{H}(U;F)$ be amply bounded. It follows easily from Definition 36.1 that there exists an open subset V' of U containing K, and a number $\rho > 0$, such that

b) $\beta\{f(x)\} \leq \rho$ for every $f \in \chi$, $x \in V'$.

From a) and b), we have

$$p(f) \leq C(V')\cdot\rho \quad \text{for every} \quad f \in \chi.$$

Therefore χ is τ_ω-bounded. Q.E.D.

PROPOSITION 36.4 If E is metrizable, the τ_o-bounded subsets of $\mathcal{H}(U;F)$, the amply bounded subsets of $\mathcal{H}(U;F)$ and the τ_ω-bounded subsets of $\mathcal{H}(U;F)$ are the same.

PROOF: It suffices to show that every τ_o-bounded set is amply bounded. Let χ be a τ_o-bounded subset of $\mathcal{H}(U;F)$. Then, for every compact $K \subset U$ and every $\beta \in CS(F)$, there exists a number $C > 0$ such that $\beta\{f(x)\} \leq c$ for every $f \in \chi$ and $x \in K$. As in the proof of Proposition 36.2, this implies that there exists an open subset V of U containing K, and a number $C_1 > 0$ such that $\beta\{f(x)\} \leq C_1$ for every $f \in \chi$, $x \in V$. Hence χ is amply bounded. Q.E.D.

REMARK 36.3: There are situations in which the τ_ω-bounded subsets of $\mathcal{H}(U;F)$ are not amply bounded (see the work of Barroso, Matos and Nachbin in [10]).

DEFINITION 36.1 A subset χ of $\mathcal{H}(U;F)$ is said to be pointwise bounded if, for every $\beta \in CS(F)$ and every $x \in U$, there exists a number $C > 0$ such that $\beta\{f(x)\} \leq C$ for every $f \in \chi$.

PROPOSITION 36.5 Let F be separated. For $\chi \subset \mathcal{H}(U;F)$ the following are equivalent:

1) $\mathcal{X}$ is amply bounded in $\mathcal{H}(U;F)$.

2) $\mathcal{X}$ is τ_o-bounded in $\mathcal{H}(U;F)$ and equicontinuous in U.

3) $\mathcal{X}$ is pointwise bounded and equicontinuous in U.

PROOF: 1) $\Rightarrow$ 2). Let $\mathcal{X}$ be amply bounded. By Proposition 36.2, $\mathcal{X}$ is τ_ω-bounded, and hence τ_o-bounded in $\mathcal{H}(U;F)$. Let $\xi \in U$ and $\beta \in CS(F)$. There exist $\alpha \in CS(E)$, and a number $c > 0$ such that

a) $B_{\alpha,1}(\xi) \subset U$;

b) If $f \in \mathcal{X}$, then $\|\frac{1}{m!}\hat{d}^m f(\xi)\|_{\alpha,\beta} \leq c$ for every $m \in \mathbb{N}$,

and

c) $\sum_{m=0}^{\infty} \frac{1}{m!}\hat{d}^m f(\xi)(x-\xi)$ converges uniformly to f in $B_{\alpha,1}(\xi)$ relative to β.

Therefore

$$\beta\{f(x)-f(\xi)\} \leq \sum_{m=1}^{\infty} \beta\{\frac{1}{m!}\hat{d}^m f(\xi)(x-\xi)\} \leq c \sum_{m=1}^{\infty} \alpha(x-\xi)^m$$

$$\leq c \sum_{m=1}^{\infty} \epsilon^m \leq c\epsilon/1-\epsilon$$

if $\alpha(x-\xi) \leq \epsilon < 1$. Hence $\mathcal{X}$ is equicontinuous at ξ.

2) $\Rightarrow$ 3) is trivial.

3) $\Rightarrow$ 1). Let $\mathcal{X}$ be pointwise bounded and equicontinuous in U. If $\xi \in U$ and $\beta \in CS(F)$, there exists a number $c > 0$ such that

a) $\beta\{f(\xi)\} \leq c$ for every $f \in \mathcal{X}$.

Furthermore, since $\mathcal{X}$ is equicontinuous, there is a neighbourhood V of ξ in U such that

b) $\beta\{f(x)-f(\xi)\} \leq 1$ for every $f \in \chi$, $x \in V$.

Combining a) and b), we have

$$\beta\{f(x)\} \leq c+1 \quad \text{for every} \quad f \in \chi, \quad x \in V,$$

and hence χ is amply bounded. Q.E.D.

EXAMPLE 36.1 Let E be a complex normed space of infinite dimension, and let E_σ denote the space E with the weak topology $\sigma(E,E')$. Let B be the closed unit ball of E'. Then $B \subset \mathcal{H}(E_\sigma)$, and B is τ_o-bounded. However, B is not amply bounded in $\mathcal{H}(E_\sigma)$; in fact,

$$\sup\{|f(x)| : f \in B, x \in V\} = +\infty$$

for every neighbourhood V of the origin in E_σ. For such a set V, there exist $f_1, f_2, \ldots, f_n \in E'$ and a number $\varepsilon > 0$ such that

$$\{x \in E : |f_i(x)| \leq \varepsilon, \quad i = 1,\ldots,n\} \subset V.$$

Since E has infinite dimension, there exists $f \in B$ which is not a linear combination of $f_1,\ldots,f_n$. Hence there exists $x_o \in E$ such that $f_1(x_o) = \ldots = f_n(x_o) = 0$, and $f(x_o) \neq 0$. Then $f_1(\lambda x_o) = \ldots = f_n(\lambda x_o) = 0$ for every $\lambda \in \mathbb{C}$, that is, $\lambda x_o \in V$ for every $\lambda \in \mathbb{C}$; but $|f(\lambda x_o)| \to \infty$ as $|\lambda| \to \infty$. Therefore B is not amply bounded. Since B is τ_o-bounded, B is not equicontinuous in E.

BIBLIOGRAPHY

<u>Note</u>: The following list of books and papers is far from complete. It contains the references quoted in the text and others that aim to arouse the reader's interest in many topics not dealt with in this book. Fairly complete bibliographies can be found in [L] and in [102]. [Q] was a landmark in the subject. References [A], [B], [C], [D], [E], [F], [G] indicate proceedings of Meetings on infinite dimensional Holomorphy and other areas, with numerous bibliographical quotations. The already-published (and more advanced) [I], [J], [K], [L], [M], [O], [R], [S], [T] and the soon-to-appear (and on the same level as this book) [H], [N], [P], all of which have different objectives from this book, are recommended as important reading material.

The reader's attention should also be drawn to [109], in which motivations suggested by different branches of Mathematics point to the necessity and importance of the study of holomorphic mappings between spaces of infinite dimension. In [U], special recommendation is made of the reading of J. Horváth's study. In the same volume, the papers by H. Upmeier and J.-F. Colombeau present suggestive aspects of the relationship between infinite dimensional Holomorphy and other branches of Mathematics and Mathematical Physics. In [27] there appears a proof of the Josefson-Nissenzweig theorem which is more accessible than the others.

[A] Proceedings on Infinite Dimensional Holomorphy (Ed.: P.L. Hayden and P.J. Suffridge). Springer-Verlag Lecture Notes in Mathematics 364 (1974).

[B] Infinite Dimensional Holomorphy and Applications (Ed.: M.C. Matos) North-Holland Pub. Co. (1977).

[C] Advances in Holomorphy (Ed.: J.A. Barroso) North-Holland Pub. Co. (1979).

[D] Functional Analysis, Holomorphy and Approximation Theory (Ed.: S. Machado) Springer-Verlag Lecture Notes in Math. 843 (1981).

[E] Functional Analysis, Holomorphy and Approximation Theory (Ed.: J.A. Barroso) North-Holland Pub. Co. (1982).

[F] Functional Analysis, Holomorphy and Approximation Theory (Ed.: G.I. Zapata) Marcel Dekker, Inc. (1983).

[G] Functional Analysis, Holomorphy and Approximation Theory (Ed.: G.I. Zapata) North-Holland Pub. Co. (1984).

[H] Chae, S.B., Holomorphy and Calculus in normed spaces. Marcel Dekker, Inc. To appear.

[I] Colombeau, J.-F., Differential Calculus and Holomorphy. North-Holland Pub. Co. (1982).

[J] Colombeau, J.-F., New Generalized Functions and Multiplication of Distributions. North-Holland Pub. Co. (1984).

[K] Coeuré, G., Analytic functions and manifolds in infinite dimensional spaces. North-Holland Pub. Co. (1974).

[L] Dineen, S., Complex Analysis in Locally Convex Spaces. North-Holland Pub. Co. (1981).

[M] Franzoni, T. and E. Vesentini, Holomorphic maps and invariant distances. North-Holland Pub. Co. (1980).

[N] Isidro, J.M. and L.L. Stachó, Holomorphic automorphism groups in Banach spaces. North-Holland Pub. Co. To appear.

[O] Mazet, P., Analytic sets in locally convex spaces. North-Holland Pub. Co. (1984).

[P] Mujica, J., Complex Analysis in Banach spaces. North-Holland Pub. Co. To appear.

[Q] Nachbin, L., Topology on spaces of holomorphic mappings. Springer-Verlag (1969).

[R] Noverraz, P., Pseudo-convexité. Convexité polynomial et domains d'holomorphie en dimension infinie. North-Holland Pub. Co. (1973).

[S] Ramis, J.P., Sous ensembles analytiques d'une variété banachique complexe. Springer-Verlag Erg. der Math. 53, (1970).

[T] Upmeier, H., Jordan C^*-Algebras and Symmetric Banach Manifolds. North-Holland Pub. Co. (1984).

[U] Advances in Mathematics and Its Applications (Ed.: J.A. Barroso) North-Holland Pub. Co. To appear.

[1] Abuabara, T., A version of the Paley-Wiener-Schwartz theorem in infinite dimensions. Advances in Holomorphy (Ed.: J.A. Barroso) North-Holland Pub. Co. (1979), 1-29.

[2] Alexander, H., Analytic functions on Banach spaces. Thesis, University of California, Berkeley (1968).

[3] Ansemil, J.M. and S. Ponte, Topologies associated with the compact-open topology on $\mathcal{H}(U)$, preprint (1980).

[4] Aragona, J., Holomorphically significant properties of spaces of holomorphic germs. Advances in Holomorphy (Ed.: J.A. Barroso) North-Holland Pub. Co. (1979), 31-46.

[5] Aron, R.M. and P.D. Berner, A Hahn-Banach extension theorem for analytic mappings. Bull. Soc. Math. France, 106 (1978), 3-24.

[6] Aron, R.M. and Carlos Herves, Weakly sequentially continuous analytic functions on a Banach space. Functional Analysis, Holomorphy and Approximation Theory (Ed.: G.I. Zapata) North-Holland Pub. Co. (1984), 23-38.

[7] Aron, R.M. and M. Schottenloher, Compact holomorphic mappings on Banach spaces and the approximation property. J. Funct. Analysis, 21, 1 (1976), 7-30.

[8] Barroso, J.A., Topologias nos espaços de aplicações holomorfas entre espaços localmente convexos. Anais Acad. Bras. de Ciências, 43 (1971), 527-546.

[9] Barroso, J.A. and L. Nachbin, Sur certaines propriétés bornologiques des espaces d'applications holomorphes. Troisième Colloque sur l'Analyse Fonctionelle, Liège 1970, C.B.R.M., Vander (1971), 47-55.

[10] Barroso, J.A., M.C. Matos and L. Nachbin, On bounded sets of holomorphic mappings. Proc. Infinite Dimensional Holomorphy (Ed.: T.L. Hayden and T.J. Suffridge) Springer-Verlag Lecture Notes in Math. 364 (1974), 31-74.

[11] Barroso, J.A., M.C. Matos and L. Nachbin, On holomorphy versus linearity in classifying locally convex spaces. Infinite Dimensional Holomorphy and Applications (Ed.: M.C. Matos) North-Holland Pub. Co. (1977), 31-74.

[12] Barroso, J.A. and L. Nachbin, Some topological properties of spaces of holomorphic mappings in infinitely many variables. Advances in Holomorphy (Ed.: J.A. Barroso) North-Holland (1979), 67-91.

[13] Barroso, J.A. and L. Nachbin, A direct sum is holomorphically bornological with the topology induced by a cartesian product. To appear in Portugaliae Math.

[14] Bierstedt, K.-D. and R. Meise, Nuclearity and the Schwartz property in the theory of holomorphic functions on metrizable locally convex spaces. Infinite Dimensional Holomorphy and Applications (Ed. M.C. Matos) North-Holland Pub. Co. (1977), 93-129.

[15] Bierstedt, K.-D. and R. Meise, Aspects of inductive limits in spaces of germs of holomorphic functions on locally convex spaces and applications to a study of $(\mathcal{H}(U),\tau_\omega)$. Advances in Holomorphy (Ed. J.A. Barroso) North-Holland Pub. Co. (1979), 111-178.

[16] Boland, J.P., Some spaces of entire and nuclearly entire functions on a Banach space. I. Jour. für die Reine und Angew. Math. 270 (1974), 38-60.

[17] Boland, J.P., Some spaces of entire and nuclearly entire functions on a Banach spaces. II. Jour. für die Reine und Angew. Math. 271 (1974), 8-27.

[18] Bremermann, H.I., Complex Convexity. T.A.M.S. 82 (1956), 17-51.

[19] Bourbaki, N. Topologie Générale, Chapitre X. Hermann (1949).

[20] Cartan, H. and P. Thullen, Zur Theorie der Singularitäten der Funktionen meherer Veränderlichen: Regularitäts und Konvergenzbereiche. Math. Annalen 106 (1932), 617-647.

[21] Chae, S.B., Holomorphic Germs on Banach Spaces. Ann. Inst. Fourier 21, 3 (1971), 107-141.

[22] Coeuré, G., Fonctions plurisousharmoniques sur les espaces vectoriels topologiques et applications a l'étude des fonctions analytiques. Ann. Inst. Fourier 20 (1970), 361-432.

[23] Colombeau, J.-F., On some various notions of infinite dimensional holomorphy. Proceedings Infinite Dimensional Holomorphy (Ed.: T.L. Hayden and T.J. Suffridge) Springer-Verlag Lecture Notes in Math. 364 (1974), 145-149.

[24] Colombeau, J.-F., Holomorphic mappings with a given asymptotic expansion at a boundary point. J. Math. Anal. and Appl., 72, 1 (1979), 274-282.

[25] Colombeau, J.-F. and B. Perrot, The Fourier-Borel transform in infinite many variables and applications. Functional Analysis, Holomorphy and Approximation Theory (Ed.: S. Machado) Springer-Verlag Lecture Notes in Mathematics 843 (1981), 163-186.

[26] Colombeau, J.-F. and Mario Matos, Convolution equations in infinite dimensions: Brief Survey, new results and proofs. Functional Analysis, Holomorphy and Approximation Theory (Ed.: J.A. Barroso) North-Holland Pub. Co. (1982), 131-178.

[27] Diestel, J., Sequence and Series in Banach Spaces. Springer-Verlag Graduate Texts in Mathematics nº 92 (1984).

[28] Dieudonné, J., Foundations of Modern Analysis. Academic Press (1960).

[29] Dineen, S., The Cartan-Thullen Theorem for Banach Spaces. Ann. Sc. Norm. Sup. Pisa, 24, 4 (1970), 667-676.

[30] Dineen, S., Holomorphy Types on a Banach Spaces. Studia Math. 39 (1972), 241-288.

[31] Dineen, S., Bounding subsets of a Banach Space. Math. Annalen 192 (1971), 61-70.

[32] Dineen, S., Holomorphic functions on (C_o,X_b)-Modules. Math. Annalen 196 (1972), 106-116.

[33] Dineen, S., Unbounded holomorphic functions on a Banach Space. J. London Math. Soc. 4, 3 (1972), 461-465.

[34] Dineen, S., Holomorphic functions on locally convex spaces, I. Locally convex topologies on $\mathcal{H}(U)$. Ann. Inst. Fourier, Grenoble, 23 (1973), 19-54.

[35] Dineen, S., Holomorphic germs on compact subsets of locally convex spaces. Functional Analysis, Holomorphy and Approximation Theory (Ed.: S. Machado) Springer-Verlag Lecture Notes in Math. 843 (1981), 247-263.

[36] Douady, A., Le problème des modules pour les sous espaces analytiques compacts d'un espace analytique donné. Ann. Inst. Fourier 16, 1 (1966), 1-95.

[37] Dwyer, T.A.W., Fourier-Borel duality and bilinear realizations of control systems. Proceedings 1976 Ames (NASA) Conference on Geometric Methods in Control (Ed.: R. Hermann and C. Martin), Math. Sci. Press, M4 (1977), 405-436.

[38] Dwyer, T.A.W., Differential equations of infinite order in vector-valued holomorphic Fock spaces. Infinite Dimensional Holomorphy and Applications (Ed.: M.C. Matos) North-Holland Pub. Co. (1977), 167-200.

[39] Fantappié, L., I funzionali analitici. Memorie della R. Academia Nazionale dei Lincei, 6, 3, 11 (1930), 453-683.

[40] Fréchet, M., Une definition fonctionelle des polynômes. Nou. Ann. Math. 9 (1909), 145-162.

[41] Fréchet, M., Les transformations ponctuelles abstraites. C.R. Acad. Sc. Paris, 180 (1925), 1816-1817.

[42] Fréchet, M., Les polynômes abstraits. Journal Math. Pures et Appl. (9), 8 (1929), 71-92.

[43] Gâteaux, R., Fonctions d'une infinité des variables indépendantes. Bull. Soc. Math. de France, 47 (1919), 70-96.

[44] Gâteaux, R., Sur diverses questions de calcul fonctionnel. Bull. Soc. Math. de France, 50 (1922), 1-37.

[45] Globevnik, J., On the ranges of analytic mappings in infinite dimensions. Advances in Holomorphy (Ed.: J.A. Barroso) North-Holland Pub. Co. (1979), 303-344.

[46] Greenfield, S.J. and N. Wallach, Automorphism groups of bounded domains in Banach spaces. T.A.M.S., 166 (1972), 45-57.

[47] Grothendieck, A., Sur certains espaces de fonctions holomorphes. Jour. für die Reine und Angew. Math. 192 (1953), Part I, 35-44, Part II, 77-95.

[48] Gruman, L. and C.O. Kiselman., Le problème de Levi dans les espaces de Banach à base. C.R. Acad. Sc. Paris 274 (1972), 821-824.

[49] Gunning, R. and H. Rossi., Analytic functions of several complex variables. Prentice Hall (1965).

[50] Gupta, C.P., Malgrange theorem for nuclearly entire functions of bounded type on a Banach space. Notas de Matemática 37, IMPA, Rio de Janeiro (1969).

[51] Harris, L., Schwarz's lemma and the maximum principle in infinite dimensional spaces. Doctoral Dissertation, Cornell University (1969).

[52] Harris, L., Schwarz-Pick systems of pseudo-metrics for domains in normed linear spaces. Advances in Holomorphy (Ed.: J.A. Barroso). North-Holland Pub. Co. (1979), 345-406.

[53] Hayden, T.L. and T.J. Suffridge, Fixed points of holomorphic mappings in Banach spaces. Proc. Am. Math. Soc., 60 (1976), 95-105.

[54] Hervé, M., Analytic continuation on Banach spaces. Several Complex Variables II (Ed.: J. Horváth) Springer-Verlag Lecture Notes in Mathematics 185 (1971), 63-75.

[55] Hewier, Y., On the Weierstrass problem in Banach spaces. Proceedings on Infinite Dimensional Holomorphy (Ed.: T.L. Hayden and T.J. Suffridge) Springer-Verlag Lecture Notes in Math., 364 (1974), 157-167.

[56] Hilbert, D., Wesen und Zieleiner Analysis der unendlich vielenunabhängigen Variabeln. Pend. del Circolo Math. di Palermo 27 (1909), 59-74.

[57] Hille, E., Methods in Classical and Functional Analysis. Addison-Wesley Pub. Co. (1972).

[58] Hirschowitz, A., Sur les suites de fonctions analytiques. Ann. Inst. Fourier, 20, 2 (1970), 403-413.

[59] Hirschowitz, A., Prolongement analytique en dimension infinie. Ann. Inst. Fourier 22 (1972), 255-292.

[60] Hörmander, L., An introduction to complex analysis in several variables. Van Nostrand (1966).

[61] Horváth, J., Topological vector spaces and distributions. Vol. I, Addison-Wesley Pub. Co. (1966).

[62] Isidro, J.M. and J.P. Mendez, Topological duality on the function space $(\mathcal{H}(U;F),\tau_\delta)$. J. Math. Anal. Appl. 67, 1 (1979), 239-248.

[63] Josefson, B., Weak sequential convergence in the dual of a Banach space does not imply norm convergence. Arkiv für Math., 13 (1975), 79-89.

[64] Kaup, W., Algebraic characterization of symmetric complex Banach manifolds. Math. Annalen 228 (1977), 39-64.

[65] Kiselman, C.O., Geometric aspects of the theory of bounds for entire functions in normed spaces. Infinite Dimensional Holomorphy and Applications (Ed.: M.C. Matos) North-Holland Pub. Co. (1977), 249-275.

[66] Köthe, G., Topological Vector Spaces, Vol. 1, Springer-Verlag (1969).

[67] Kramm, B., Analytische Struktur in Spektren ein Zugang über die ∞-dimensionale Holomorphie. J. Funct. Analysis, 37, 3 (1980), 249-270.

[68] Krée, P., Théorie de la mesure et holomorphie en dimension infinie. Séminaire Pierre Lelong 1975/76. Springer-Verlag Lecture Notes in Math., 578 (1977), 44-70.

[69] Krée, P., Holomorphie et théorie des distributions en dimension infinie. Infinite Dimensional Holomorphy and Applications (Ed.: M.C. Matos) North-Holland Pub. Co. (1977), 277-296.

[70] Lelong, P., Sur les fonctions plurisousharmoniques dans les espaces vectoriels topologiques et une extension du théorème de Banach-Steinhaus aux families d'applications polynomiales. Troisième Coll. Anal. Fonct., Liège 1970, Vander, Belgium (1971), 21-45.

[71] Lelong, P., Sur l'application exponentielle dans l'espace des fonctions entières. Infinite Dimensional Holomorphy and Applications (Ed.: M.C. Matos) North-Holland Pub. Co. (1977), 297-311.

[72] Lelong, P., A class of Fréchet complex spaces in which the bounded sets are ℂ-polar sets. Functional Analysis, Holomorphy and Approximation Theory (Ed.: J.A. Barroso) North-Holland Pub. Co. (1982), 255-272.

[73] Lelong, P., Two equivalent definitions of the density numbers for a plurisubharmonic function in a topological vector space. Functional Analysis, Holomorphy and Approximation Theory (Ed.: G.I. Zapata) North-Holland Pub. Co. (1984), 113-132.

[74] Ligocka, E., A local factorization of analytic functions and its applications. Studia Math. 47 (1973), 239-252.

[75] Martineau, A., Sur les fonctionnelles analytiques et la transformation de Fourier-Borel. J. Anal. Math., 11 (1963), 1-164.

[76] Martineau, A., Oeuvres de André Martineau (Ed.: A. Hirschowitz) Éditions du CNRS, Paris (1977).

[77] Matos, M.C., On the Cartan-Thullen theorem for some subalgebras of holomorphic functions in a locally convex space. J. für die Reine und Angew. Math. 270 (1974), 7-14.

[78] Matos, M.C., Convolution equations in spaces of uniform nuclear entire functions. Functional Analysis, Holomorphy and Approximation Theory (Ed.: G.I. Zapata) Marcel Dekker, Inc. (1983), 207-231.

[79] Matos, M.C., On the Fourier-Borel transformation and spaces of entire functions in a normed space. Functional Analysis, Holomorphy and Approximation Theory (Ed.: G.I. Zapata) North-Holland Pub. Co. (1984), 139-169.

[80] Mazet, P., Généralization des notions d'anneau noethérien et d'anneau de Cohen-Macauley. Applications à la géometrie de dimension infinie. Fonctions analytiques de Plusieurs Variables et Analyse Complexe. Colloq. Int. du C.N.R.S., Paris 1972. Agora Math. Gauthier-Villars (1974), 131-140.

[81] Meise, R. and D. Vogt, The symmetric tensor algebra of a locally convex space and entire functions (Preprint).

[82] Meise, R. and D. Vogt, Counterexamples for holomorphic functions on nuclear Fréchet space. (Preprint).

[83] Meise, R. and D. Vogt, An interpretation of τ_ω and τ_δ as normal topologies of sequence spaces. Functional Analysis, Holomorphy and Approximation Theory (Ed.: J.A. Barroso) North-Holland Pub. Co. (1982), 273-285.

[84] Michael, E.A., Locally multiplicatively-convex topological algebras. Memoirs A.M.S., 11 (1952).

[85] Michal, A.D., Le Calcul Différentiel dans les espaces de Banach. Gauthier-Villars (1958).

[86] Moraes, L.A., Tipos de holomorfia e abertos de Runge. Partes I e II. Anais Acad. Bras. de Ciências 50 (3) (1978), 277-293 e 51 (1) (1978), 5-25.

[87] Moraes, L.A., Holomorphic functions on holomorphic inductive limits and on the strong duals of strict inductive limits. Functional Analysis, Holomorphy and Approximation Theory (Ed.: G.I. Zapata) North-Holland Pub. Co. (1984), 297-310.

[88] Mujica, J., Spaces of germs of holomorphic functions. Thesis, University of Rochester (1974). Studies in Analysis, Advances in Math. Suppl. Studies, vol. 4 (Ed. G.C. Rota) Academic Press (1979), 1-41.

[89] Mujica, J., A new topology on the space of germs of holomorphic functions. (Preprint).

[90] Nachbin, L., Lectures on the theory of idstributions. Textos de Matemática nº 15, Universidade Federal de Recife, Pernambuco, Brasil (1964). Reprinted by University Microfilms International, USA (1980).

[91] Nachbin, L., On spaces of holomorphic functions of a given type. Functional Analysis. Irvine, 1966. (Ed.: B.R. Gelbaum). Academic Press (1967), 55-70.

[92] Nachbin, L., On the topology of the space of all holomorphic functions on a given open subset. Indagationes Mathematicae, 29 (1967), 366-368.

[93] Nachbin, L., Sur les espaces vectoriels topologiques d'applications continues. C.R. Acad. Sc. Paris, 271 (1970), 596-598.

[94] Nachbin, L., Convolution operators in spaces of nuclearly entire functions on a Banach space. Functional Analysis and Related Fields, (Ed.: F.E. Browder) Springer-Verlag (1970), 167-171.

[95] Nachbin, L., Concerning holomorphy types for Banach spaces. Intern. Coll. Nuclear Spaces and Ideals in Operator Algebras. Studia Math., 38 (1970), 407-412.

[96] Nachbin, L., Concerning spaces of holomorphic mappings. Publications of Depart. of Math., Rutgers University (1970).

[97] Nachbin, L., Uniformité d'holomorphie et type exponential. Séminaire Pierre Lelong 1970. Springer-Verlag Lecture Notes in Math. 205 (1971), 216-224.

[98] Nachbin, L., Sur quelques aspects récents d'holomorphie en dimension infinie. Séminaire Goulaouic-Schwartz, Paris 1971/72, 18, 1-9.

[99] Nachbin, L., Weak holomorphy, Part I and II. Unpublished manuscript (1972).

[100] Nachbin, L. and S. Dineen, Entire functions of exponential type bounded on the real axis and Fourier transforms of distributions with bounded support. Israel J. Math. 13 (1972), 321-326.

[101] Nachbin, L., On vector-valued versus scalar-valued holomorphic continuation. Indagationes Math., 35, 4 (1973), 352-354.

[102] Nachbin, L., Recent developments in infinite dimensional holomorphy. B.A.M.S. 79 (1973), 625-640.

[103] Nachbin, L., Limites et perturbations des applications holomorphes. Fonctions analytiques de Plusieurs Variables et Analyse Complexe, Colloq. Int. du C.N.R.S., Paris 1972. Agora Math. Gauthier-Villars (1974), 131-140.

[104] Nachbin, L., A glimpse at infinite dimensional holomorphy. Proc. on Infinite Dimensional Holomorphy (Ed.: T.L. Hayden and T.J. Suffridge) Springer-Verlag Lecture Notes in Math. 364 (1974), 69-79.

[105] Nachbin, L., Perturbation of holomorphic mappings revisited. Rend. di Mat., (2), Vol. 8 (1975), 337-344.

[106] Nachbin, L., Some holomorphically signficant properties of locally convex spaces. Funct. Analysis (Ed.: D.J. Figueiredo) Marcel Dekker, Inc. (1976).

[107] Nachbin, L., Analogie entre l'holomorphie et la linéarité. Séminaire Paul Krée. Équations aux Dérivées Partielles en Dimension Infinie, 1977/78, nº 1.

[108] Nachbin, L., Some problems in the application of functional analysis to holomorphy. Advances in Holomorphy (Ed. J.A. Barroso) North-Holland Pub. Co. (1979), 577-583.

[109] Nachbin, L., Why holomorphy in infinite dimensions. L'Enseignement Mathématique 26 (1980), 257-269.

[110] Nachbin, L. and M.C. Matos, Silva holomorphy types. Functional Analysis, Holomorphy and Approximation Theory (Ed.: S. Machado) Springer-Verlag Lecture Notes in Math. 843 (1981), 437-487.

[111] Nachbin, L. and M.C. Matos, Entire functions on locally convex spaces and convolution operators. Compositio Math. 44 (1981), 145-181.

[112] Nissenzweig, A., w*-sequential convergence. Israel J. Math. 22 (1975), 266-272.

[113] Noverraz, Ph., Fonctions plurisousharmoniques et analytiques dans les espaces vectoriels topologiques. Ann. Inst. Fourier 19, 2 (1969), 419-493.

[114] Noverraz, Ph., Sur le théorème de Cartan-Thullen-Oka en dimension infinie. Séminaire Pierre Lelong 1971/72. Springer-Verlag Lecture Notes in Math. 332 (1973), 59-68.

[115] Noverraz, Ph., Le problème de Levi en dimension infinie. Bull. Société Math. de France, Mémoire 46 (1976), 73-82.

[116] Oka, K., Sur les fonctions analytiques de plusieurs variable, VI. Domaines pseudo-convexes. Tohoku Math. J., 49 (1942), 15-52.

[117] Oka, K., Sur les fonctions analytiques de plusieurs variables complexes, IX. Domains finis sans point critique intérieur. Japan J. Math. 23 (1953), 97-155.

[118] Pisanelli, D., Sur la (LF)-analyticité. Analyse Fonctionelle et Applications (Ed.: L. Nachbin) Herman Act. Sc. et Ind. (1975), 215-224.

[119] Pisanelli, D., Applications analytiques en dimension infinie. Bull. Société Math. de France 96 (1976), 181-191.

[120] Raboin, P., Le problème du $\bar{\partial}$ sur un espace de Hilbert. Bull. Société Math. de France, 107 (1979), 225-240.

[121] Ramis, J.P., Sous ensemble analytiques d'une variété analytique banachique. Séminaire Pierre Lelong 1967/68. Springer-Verlag Lecture Notes in Math., 71 (1968), 140-164.

[122] Rickart, C.E., Analytic functions of an infinite number of complex variables. Duke Math. Journal 36 (1969), 581-597.

[123] Rickart, C.E., Natural functions algebras. Springer-Verlag Universitext (1979).

[124] Ryan, R.A., Applications of topological tensor products to infinite dimensional holomorphy. Thesis. Trinity College Dublin (1980).

[125] Schottenloher, M., Analytische Fortsetzung in Banachräumen. Dissertation, Munich (1971).

[126] Schottenloher, M., ε-product and continuation of analytic mappings. C.R. du Colloque d'Analyse Fonctionnelle et Applications (Ed.: L. Nachbin) Rio de Janeiro 1972, Hermann (1974), 261-270.

[127] Schottenloher, M., The Levi problem for domains spread over locally convex spaces with a finite dimensional Schauder decomposition. Ann. Inst. Fourier 26, 4 (1976), 207-237.

[128] Sebastião e Silva, J., As funções analíticas e a Análise Funcional. Port. Math., 9 (1950), 1-130.

[129] Sebastião e Silva, J., Le calcul différentiel et intégral dans les espaces localement convexes, réels ou complexes, I. Atti Acad. Lincei Rend., 20 (1956), 743-750.

[130] Silva Dias, C.L. da, Espaços vetoriais topológicos e suas aplicações nos espaços de funcionais analíticos. Bol. Soc. Mat. São Paulo 5 (1950), 1-58.

[131] Soraggi, R.L., Bounded sets in spaces of holomorphic germs. Advances in Holomorphy (Ed.: J.A. Barroso) North-Holland Pub. Co. (1979), 745-766.

[132] Soraggi, R.L., Holomorphic germs on certain locally convex spaces (1980). (Preprint).

[133] Stevenson, J.O., Holomorphy of composition. Infinite Dimensional Holomorphy and Applications (Ed.: M.C. Matos) North-Holland Pub. Co. (1977), 397-427.

[134] Taylor, A.E., On the properties of analytic functions in abstract space. Math. Annalen, 115 (1938), 466-484.

[135] Taylor, A.E., Historical notes on analyticity as a concept in functional analysis. Problems in Analysis (Ed.: R.C. Gunning) Princeton Math. Series, Vol. 31 (1970), 325-343.

[136] Taylor, A.E., Notes on the history of the uses of analyticity in operator theory. Am. Math. Monthly 78 (1971), 331-342.

[137] Thorp, E. and R. Whitley, The strong maximum modulus theorem for analytic functions into a Banach space. P.A.M.S. 18 (1967), 640-646.

[138] Upmeier, H., A holomorphic characterization of C*-algebras. Functional Analysis, Holomorphy and Approximation Theory (Ed.: G.I. Zapata) North-Holland Pub. Co. (1984), 427-467.

[139] Vesentini, E., Maximum theorems for vector-valued holomorphic functions. Rend. Sem. Mat. Fis. Milan 40 (1970), 23-55.

[140] Vigué, J.P., Le groupe des automorphismes analytiques d'un domaine borné d'un espace de Banach complexe. Ann. Sc. Ec. Norm. Sup., 4, 9 (1976), 203-282.

[141] Vigué, J.P., Sur la convexité des domaines bornés circlés homogènes. Séminaire Pierre Lelong-H. Skoda, 1978/79. Springer-Verlag Lecture Notes in Math. 822 (1980), 317-331.

[142] Volterra, V., Sopra le funzioni che dependone da altre funzioni. Nota I. Rend. Accad. Lincei, Series 4, Vol. 3 (1887), 97-105.

[143] Volterra, V., Sopra le funzioni che dependone da altre funzioni. Nota II. Rend. Accad. Lincei, Series 4, Vol. 3 (1887), 141-146.

[144] Volterra, V., Sopra le funzioni che dependone da altre funzioni. Nota III. Rend. Accad. Lincei, Series 4, Vol. 3 (1887), 153-158.

[145] Volterra, V., Sopra le funzioni dependenti da linee. Nota I. Rend. Accad. Lincei, Series 4, Vol. 3 (1887), 225-230.

[146] Volterra, V., Sopra le funzioni dependenti da linee. Nota II. Rend. Accad. Lincei, Series 4, vol. 3 (1887), 274-281.

[147] Waelbroeck, L., The nuclearity of $\mathcal{O}(U)$. Infinite Dimensional Holomorphy and Applications (Ed.: M.C. Matos) North-Holland Pub. Co. (1977), 425-436.

[148] Wanderley, A.J.M., Germes de aplicações holomorfas em espaços localmente convexos. Tese, Universidade Federal do Rio de Janeiro (1974).

[149] Zorn, M., Characterization of analytic functions in Banach spaces. Annals of Math. 46 (1945), 585-593.

AN INDEX OF DEFINITIONS

Amply bounded mapping 163
Amply bounded set of mappings 218
Balanced set 45
Bounded topologies on $\mathcal{L}_s(^mE;F)$ and $\mathcal{P}(^mE;F)$ 171
Coefficients of a power series 17
Compact-open topology on C(U;F) (τ_o-topology) 235
Compact topologies on $\mathcal{L}_s(^mE;F)$ and $\mathcal{P}(^mE;F)$ 171
Continuous m-homogeneous polynomial 4
Convergent (uniformly convergent) power series 17
Degree of a polynomial 163
Differentially stable space 233
Differentials of a holomorphic mapping 26
Domains of $\mathcal{H}_b$-holomorphy 133
Entire mappings 28

Space of entire mappings $\mathcal{H}(E;F)$ 28

Entire mappings of bounded type 119

Space of entire mappings of bounded type $\mathcal{H}_b(E;F)$.. 119

Finite topologies on $\mathcal{L}_s(^mE;F)$ and $\mathcal{P}(^mE;F)$ 171
Finitely holomorphic mapping 69,245

Space of finitely holomorphic mappings, $\mathcal{H}_{fh}(U;F)$.. 69

Formal power series 173

Space of formal power series, $F_a[[E]]$ 173

Space of continuous formal power series, F[[E]]. 173

G-holomorphic (Gâteaux holomorphic) mapping 75
$\mathcal{H}$-bounding set 202
$\mathcal{H}_b$-domain of existence 150
$\mathcal{H}_b$-holomorphic extension 131
$\mathcal{H}_b$-holomorphic prolongation 132
$\mathcal{H}_b$-holomorphically convex open set 142
$\mathcal{H}_b$-hull of a subset X of U 139
$\mathcal{H}_b$-regular holomorphic mapping 150

$\mathcal{H}_b$-singular point of a holomorphic mapping 150
Holomorphic mapping 25,177
Space of holomorphic mappings $\mathcal{H}(U;F)$ 25,177
Holomorphic mapping of bounded type 81
Space of holomorphic mappings of bounded type, $\mathcal{H}_b(U;F)$ 81
Holomorphic mappings between topological vector spaces which are not locally convex 180
Limit topologies on $\mathcal{L}_s(^mE;F)$ and $\mathcal{P}(^mE;F)$ 170,171
Locally bounded mapping 48
m-homogeneous polynomial 4,159
Space of m-homogeneous polynomials, $\mathcal{P}_a(^mE;F)$. 4,159
Nachbin topology (τ_ω-topology) 85,263
Natural topology on $\mathcal{H}_b(U;F)$ 81
Natural topology on $\mathcal{H}_I(U;F)$ 97
Norm on $\mathcal{L}(E_1,\ldots,E_m;F)$ 2
Norm on $\mathcal{P}(^mE;F)$ 5
Polarization formula 6,160
Polynomial 13
Space of polynomials, $\mathcal{P}_a(E;F)$ 161
Power series 17
Proper $\mathcal{H}_b$-holomorphic extension 131
Proper $\mathcal{H}_b$-holomorphic prolongation 132
Radius of boundedness of a holomorphic mapping at a point 48
Radius of convergence (radius of uniform convergence) of a power series 17
Schauder basis 108
Seminorm on $\mathcal{L}_s(^mE;F)$ 167
Seminorm on $\mathcal{P}(^mE;F)$ 167
Seminorm ported by a compact 85,263
Separately holomorphic mapping 76
Set of all mappings from E into F, $\mathfrak{F}(E;F)$ 11
Silva spaces 260
Space of bounded holomorphic mappings, $\mathcal{H}_B(U;F)$ 83
Space of continuous formal power series, F[[E]] ... 173

Space of continuous m-homogeneous polynomials, $\mathcal{P}(^mE;F)$. 159
Space of continuous m-linear mappings
$\mathcal{L}(E_1,\ldots,E_m;F)$, $\mathcal{L}(^mE;F)$ 1,2,3,156
Space of continuous polynomials, $\mathcal{P}(E;F)$ 161
Space of continuous symmetric m-linear
mappings, $\mathcal{L}_s(^mE;F)$ 3,156
Space of entire mappings, $\mathcal{H}(E;F)$ 28
Space of finitely holomorphic mappings, $\mathcal{H}_{fh}(U;F)$.. 69
Space of formal power series, $F_a[[E]]$ 173
Space of holomorphic mappings, $\mathcal{H}(U;F)$ 25,177
Space of holomorphic mappings of bounded type,
$\mathcal{H}_b(U;F)$, $\mathcal{H}_b(E;F)$ 81,119
Space of holomorphic mappings which are bounded
on each set of an open contable cover of U,
$\mathcal{H}_I(U;F)$ 97
Space of m-homogeneous polynomials, $\mathcal{P}_a(^mE;F)$ 159
Space of m-linear mappings, $\mathcal{L}_a(E_1,\ldots,E_m;F)$,
$\mathcal{L}_a(^mE;F)$ 1,155
Space of polynomials, $\mathcal{P}_a(E;F)$ 161
Space of symmetric m-linear mappings, $\mathcal{L}_{as}(^mE;F)$ .. 3,155
Subconvex set 136
Symmetrization of a m-linear mapping 3,155
Taylor coefficient of a holomorphic mapping 177
Taylor polynomial of order m of a holomorphic
mapping 178
Taylor series of a holomorphic mapping
at a point 25,178
Topology τ_ℓ on $\mathcal{H}(U;F)$ 270
Topology τ_o in $C(U;F)$ 38,235
Topology τ_o in $\mathcal{H}(U;F)$ induced by the
compact-open topology on $\mathcal{H}(U;F)$ 84,263
Topology τ_∞ on $\mathcal{H}(U;F)$ 84
Topology τ_m on $\mathcal{H}(U;F)$ 84
Topology τ_δ on $\mathcal{H}(U;F)$ 100
Topology τ_ω 85,263
U-bounded set 81

AUTHOR INDEX

Banach, S. 2
Barroso, J.A. 276
Bourbaki, N. 39
Cartan, H. 126,139,146,150,152
Cauchy, A.L. 18,31,32,35,209,210,212,229
Coeuré, G. 100
Dieudonné, J. 31
Dineen, S. 108,109,154
Fréchet, M. 81
Gateaux, R. 75
Hadamard, J. 18
Hartogs, F. 77
Hille, E. 31
Hirschowitz, A. 206
Josefson, B. 122,204
Martineau, A. 85
Matos, M.C. 276
Montel, P. 151
Nachbin, L. 26,85,100,276
Newton, I. 157
Nissenzweig, A. 102,204
Schauder, J. 108
Silva, J.S. 260
Steinhaus, H. 66
Taylor, B. 25,215
Thorp, E. 116
Thullen, P. 126,139,146,150,152
Vesentini, E. 116
Witley, R. 116